X-RAYS:
The First Hundred Years

X-RAYS:
The First Hundred Years

Edited by
Alan Michette and Sławka Pfauntsch
King's College London, UK

JOHN WILEY & SONS

Chichester • New York • Brisbane • Toronto • Singapore

Other Wiley Editorial Offices

John Wiley & Sons, Inc., 605 Third Avenue,
New York, NY 10158-0012, USA

Jacaranda Wiley Ltd, 33 Park Road, Milton,
Queensland 4064, Australia

John Wiley & Sons (Canada) Ltd, 22 Worcester Road,
Rexdale, Ontario M9W 1L1, Canada

John Wiley & Sons (Asia) Pte Ltd, 2 Clementi Loop #02-01,
Jin Xing Distripark, Singapore 0512

Library of Congress Cataloging-in-Publication Data

British Library Cataloguing in Publication Data

A catalogue record for this book is available from the British Library

ISBN 0-471-96502-2

Reproduced from author's CRC
Printed and bound in Great Britain by Biddles Ltd, Guildford and King's Lynn
This book is printed on acid-free paper responsibly manufactured from sustainable forestation,
for which at least two trees are planted for each one used for paper production.

Contents

Preface ix

Contributors xiii

The First Hundred Years 1
Albert Franks

The Discovery of X-Rays 1
Production of X-Rays 4
Applications of X-Rays 6
X-Ray Optics 14
Summary 18
Bibliography 19
References 19

The Early History of X-Rays in Medicine for Therapy and
Diagnosis 21
Richard Mould

The Discovery of X-Rays 22
Public Reaction 25
Radiography 26
Herbert Jackson of King's College 30
Fluoroscopy 31
Photofluoroscopy, Cine-Radiography and Arteriography 33
Radiation Protection 34
X-Ray Therapy 36
Radiation Units 38
X-Ray Quality 40
Summary 40
Bibliography 40

X-Ray Microscopy... **43**
William Nixon

Contact Microradiography.. 43
Reflection X-Ray Microscopy.. 44
Projection X-Ray Microscopy 44
X-Ray Microscopy using Zone Plates 48
Lensless X-Ray Microscopy... 57
Summary... 58
Bibliography 58
References 59

X-Ray Microanalysis... **61**
James Long

X-Ray Spectra.. 61
Chemical Analysis with X-Rays..................................... 65
X-Ray Fluorescence Analysis of Bulk Samples............. 66
Localised Analysis and the Electron Microprobe 75
Beam Scanning 79
Energy Dispersive Spectrometers.................................. 84
Microanalysis in the Transmission Electron Microscope 87
Localised Trace Element Analysis with X-Ray Excitation............. 88
Localised Trace Element Analysis with Proton Excitation............. 90
Current Activity.. 95
References 96

X-Ray Diffraction... **101**
Watson Fuller

Beyond the Geometrical Shadow 101
The Phase Problem 102
Solutions to the Phase Problem 107
Highlighting Regularity... 110
Functional Advantages of Symmetry in Biological Systems 110
The Special Case of Mirror Symmetry........................... 113
Characterising Hierarchies 114
Molecular Structures 114
Macromolecular Assemblies 118
Structure in Context... 120

Relationships Between Structure and Function.................................... 120
Specificity in Partially Ordered Systems.. 122
Following Change... 124
Modelling and Design 127
References 128

Synchrotron Radiation **131**
Ian Munro

Fundamental Properties of Synchrotron Radiation 135
The Evolution of 'Photon Factories' 137
The Ultimate Prospects for Synchrotron Radiation Sources............... 140
Coherence 143
Why is Synchrotron Radiation Important?.. 146
Biological Science 147
Materials and Surface Sciences... 149
Molecular Science 151
Links with Industry.. 152
The Future of Synchrotron Radiation.. 152
Bibliography 154

X-Ray Lithography... **155**
Alistair Smith

Optical Lithography... 158
X-Ray Lithography... 161
The Helios Design 164
Helios Performance 170
References 172

X-Ray Astronomy... **175**
Ken Pounds

The Sun as an X-Ray Source... 175
Cosmic X-Ray Astronomy—the Early Days....................................... 177
The Satellite Era 179
The Modern Era.. 183
The Scientific Impact of X-Ray Astronomy....................................... 186
The Future 190
References 190

The Development and Applications of Laser Produced Plasma X-Ray Sources 193
Ciaran Lewis

Incoherent X-Rays 193
Spectral Features 194
Applications 198
Coherent X-Rays 206
Development of Collision Pumped X-Ray Lasers 211
Other Approaches to X-Ray Lasing 219
Applications of X-Ray Lasers 220
Bbibliography 223

How Lasers Generate Bright Sources of X-Rays 225
Mike Key

Intensity in Radiation Sources 228
New High Power and Ultrashort Pulse Lasers 228
XUV Harmonic Generation 231
Soft X-Ray Lasers 234
Laser Generated Electron Beams and Novel X-Ray Sources 236
Comparison of XUV and X-Ray Sources 237
An Application of XUV Lasers to Radiography 238
Conclusions 240
References 240

The Next Hundred Years 243
Andrew Miller

X-Ray Sources 244
Detectors 245
Applications 245
Summary 248
References 248

Glossary 249

Index 259

Preface

This book grew from a great discovery and a small idea. Since 1991 the Centre for X-Ray Science in the Department of Physics at King's College London has held regular weekly one hour seminars on Wednesday afternoons, covering all aspects of research into and using x-rays. When, in late 1994, we began to plan the 1995 seminar programme, we realised that 8 November 1995, to the day the exact one hundredth anniversary of the discovery of x-rays by Wilhelm Röntgen, fortuitously fell on a Wednesday. The seminar for that day had to celebrate the centenary, but what topic should we choose from the many in which x-rays play an essential rôle?

Everybody, of course, is familiar with the medical and dental uses of x-rays and with security checks at airports, and many people are aware of other applications such as x-ray astronomy. But these are just the tip of the iceberg and, as we discussed possible seminar topics and speakers, it soon became clear that one hour would not be enough. A little later, it became obvious that half a day would also not allow enough time to cover the material adequately, and so the idea of a one day meeting with several talks was born.

The book, then, is based on the presentations made at that meeting by international scientists each of whom is an expert in his area of x-ray science, ranging from the microworld to distant astronomical objects, with the aim of giving an historical flavour as well as an up to date picture. Even so, it would never be possible in the time available for the meeting or the space available in the book to give a fully comprehensive coverage, and so some equally important topics will, inevitably and unfortunately, be missing. From the outset, we tried to ensure that the presentations at the meeting, and the written versions included here, would be given in as nontechnical a way as possible, so that they would appeal to a broad audience. Given the rather esoteric nature of some of the work described, technical terms have, of course, crept in, and so we have included a glossary to explain some of these. The definitions in the glossary are, of necessity, brief and the reader is referred to the bibliographies provided by the authors for fuller explanations.

Shortly after Röntgen's momentous discovery scientists all over the World were closely involved with x-rays and this has continued to the present day. The Physics Department at King's has a long standing reputation for high quality research, resulting in four Nobel Prizes. Those awarded to Charles Barkla (1917) for "his discovery of the characteristic Röntgen radiation of the elements" and Maurice Wilkins (1962) for "discoveries concerning the molecular structure of nuclear acids and its significance for information transfer in living material" represent two highlights of the x-ray work at King's, but are not the whole story.

Well before Barkla's work, Herbert Jackson, then a chemistry demonstrator at King's and later a professor, made one of the first major advances in x-ray tube design, incorporating a curved cathode to focus the x-ray beam. Much later, in 1976, Michael Hart was appointed to the Wheatstone Chair of Physics and pioneered work in x-ray interferometry which allowed precise measurements of the optical properties of materials at x-ray energies. Charles Wheatstone was professor of experimental philosophy at King's from 1834 until his death in 1875; incidentally, he did not invent the resistance bridge named after him, but he did invent the accordion and the electric telegraph. The other established chair in the department, for theoretical physics, is named after James Clerk Maxwell who, while professor of physics at King's from 1860–65, formulated his famous equations linking electricity and magnetism, laying the foundations for the theory of electromagnetic radiation and thus of the nature of x-rays.

One of the current topics of revived interest in x-ray research is the development and application of x-ray microscopes. This modern renaissance of x-ray microscopy had to await high brilliance sources, such as synchrotrons, and the refinement of techniques for making high resolution x-ray optics such as zone plates. X-ray microscopy has the advantage over optical microscopy of allowing much better resolution in images of specimens, which may be studied in their natural environment. Electron microscopy has led to many major advances in our understanding of biology, but requires careful specimen preparation. X-rays interact differently to electrons and so they can provide new and different information. Members of the Centre for X-Ray Science at King's are active in developing and using x-ray microscopy. Applications include the quantitative mapping of biologically important elements such as calcium, pioneered by Chris Buckley, and imaging of systems important in materials science, such as zeolites and silica gels, being carried out by Graeme Morrison. The main research of the editors of *X-Rays: The First Hundred Years* is in developing laboratory based x-ray microscopes using laser plasma sources. We are also investigating other uses of x-ray optics, such as the formation of microprobes to study the radiation response of biological cells, important in research into radiation induced cancers.

The work at King's, along with much of the other research described in the following pages, will, we hope, contribute to the continuing relevance of x-rays to mankind in the next hundred years and beyond.

We are grateful to all the authors for giving up their valuable time, enabling us to provide such a wide coverage; any errors which may have crept in are the sole responsibility of the editors. We also wish to thank our colleagues in the Centre for X-Ray Science, especially Peter Anastasi, for their encouragement and support, and the staff at John Wiley and Sons for their patience and assistance during the editorial process.

Alan Michette and Sławka Pfauntsch
Centre for X-Ray Science
King's College London

Contributors

Professor Albert Franks CBE, Centre of Mechanical & Optical Technology, National Physical Laboratory, Teddington, Middlesex, TW11 0LW, England

Professor Watson Fuller, Head of Physics Department, Keele University, Keele, Staffordshire, ST5 5BG, England

Professor Mike Key, Head of Central Laser Facility, CLRC Rutherford Appleton Laboratory, Oxfordshire, OX11 0QX, England.

Dr William Nixon, Peterhouse, Cambridge University, Cambridge, CB2 1RD, England

Professor Ciaran Lewis, Department of Pure & Applied Physics, The Queen's University of Belfast, Belfast BT7 1NN, Northern Ireland

Dr James Long, Department of Earth Sciences, Bullard Laboratories, Madingley Road, Cambridge, CB3 OEZ., England

Professor Andrew Miller, Principal & Vice Chancellor, University of Stirling, Stirling, FK9 4LA, Scotland

Dr Richard Mould, Scientific Consultant, 41 Ewhurst Avenue, Sanderstead, South Croydon, Surrey, CR2 0DH, England

Professor Ian Munro, Head of Synchrotron Radiation Programme, CLRC Daresbury Laboratory, Warrington WA4 4AD, England

Professor Ken Pounds CBE FRS, Chief Executive & Deputy Chairman, Particle Physics & Astronomy Research Council, Polaris House, North Star Avenue, Swindon, Wiltshire, SN2 1SZ, England

Dr Alistair Smith, Executive Manager, Plasma & Accelerator Technologies, Oxford Instruments, Eynsham, Oxfordshire, OX8 1TL, England

The First Hundred Years

Albert Franks
National Physical Laboratory

T oday, much effort is expended in furthering the rapid exploitation of the results of fundamental research. We might suppose
that a very different state of affairs existed in the more leisurely age
of a hundred years ago. However, it would be difficult even today to
emulate the rapidity with which the discovery of x-rays was exploited. Wilhelm Conrad Röntgen (1845–1923) discovered x-rays
on 8 November 1895 in Würzburg in Germany, and in May 1896 the
first x-ray journal, *Archives of Clinical Skiagraphy*, was published in
Great Britain. In the same month, x-rays were first used on the
battlefield, in the Italian-Ethiopian campaign. The fact that x-rays
could be used to photograph bones inside the body and to locate
foreign bodies such as bullets or swallowed coins captured the
public imagination immediately. The biological effects of x-rays
were not appreciated at first and many of the scientists who began to
study them took risks that nowadays would be carefully guarded
against—and they suffered accordingly. Six years after the discovery, Röntgen was awarded a Nobel Prize, the first one in physics.

The Discovery of X-Rays

The discovery of x-rays was the culmination of more than a century
of research on electrical discharges in evacuated vessels. Without
doubt, x-rays had been generated many times before their discovery,
particularly in the 1880s when experiments with the cathode ray
tubes of Sir William Crookes were very much in vogue. Sir William
himself unsuccessfully sought the cause of the repeated and unac-
countable fogging of photographic plates stored near his cathode ray
tubes. Röntgen was using such a tube, shown in figure 1, covered in

X-RAYS: The First Hundred Years
Edited by Alan Michette and Slawka Pfauntsch © Crown Copyright 1996. Reproduced by
permission of the Controller of HMSO. Published by John Wiley & Sons Ltd

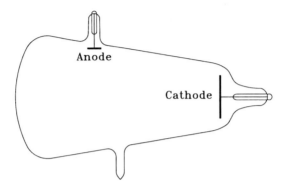

Figure 1. A diagram of a gas-filled cathode ray tube of the type used by Röntgen in the discovery of x-rays: gas pressure 0.04–0.1 Pa; anode voltage 30–100 kV.

black paper, to study the fluorescence produced when cathode rays struck the glass wall of the tube. In the darkened room he saw a brilliantly glowing screen of barium platinocyanide placed some distance away and deduced that invisible radiation was passing through the air from the tube to the screen. He subsequently showed that the radiation would pass through a piece of aluminium 15 mm thick although with much reduced intensity. He called the radiation 'x-rays', a name that has been almost universally adopted, although in Germany they are known as Röntgenstrahlen in his honour.

It was soon demonstrated that the production of x-rays was associated with the stoppage of cathode rays, or electrons as they soon became to be known, when they struck a target. But what were these mysterious x-rays? The problem of determining their nature was vigorously pursued by many investigators and an early disco-very provided a method of detecting them in addition to fluorescence and to blackening of photographic film. They were found to ionise gases, i.e., to make gases conductors of electricity. The ionisation current thus produced gave a measure of the intensity of the x-rays, but also provided a problem for the physicists trying to understand their nature. Early attempts to reflect, refract or diffract x-rays proved unsuccessful or inconclusive and this, together with their ionising power, seemed to imply that they were corpuscular in nature. Yet they travelled in straight lines and, unlike electrons or α-particles, they were not affected by electric or magnetic fields.

They could also be scattered by, for example, a block of paraffin wax, and the rays scattered in certain directions were polarised, since a second similar block could scatter them only in particular directions relative to the incident beam. These facts seemed to prove that x-rays were wavelike in nature. Röntgen assumed that they were longitudinal waves similar to sound waves, but of a much higher frequency. As he was unable to demonstrate the familiar optical phenomena of reflection, refraction, interference and diffraction, he concluded that x-rays differed fundamentally from visible or ultraviolet light and thus were not a form of electromagnetic radiation.

Some of the other early investigators, notably J J Thomson, initially agreed with Röntgen's assumption about the longitudinal wave nature of x-rays, but others speculated that the radiation was electromagnetic. It was quickly realised that if that was the case, the wavelength of the radiation would have to be very short, compared with that of visible light, and great care would have to be taken to detect any of the familiar optical phenomena. During the next fifteen years numerous and increasingly more sophisticated attempts were made to reflect, refract and diffract x-rays. Until 1912, all the experiments failed completely to show any effects that could not be attributed to instrumental causes. In that year Koch made careful photometric measurements of photographic plates, produced a few years earlier by Walter and Pohl, that had been exposed to x-rays through a wedge shaped slit a few micrometres across at its widest part. Sommerfeld calculated that the broadening of the beam that had passed through the narrowest part of the slit was caused by diffraction of radiation of wavelength between 0.01 and 0.1 nm. Because of the experimental difficulties, the results were not very convincing, and it was not until the 1920s that well defined x-ray diffraction patterns were observed which agreed accurately with the usual theory of the phenomenon as given in the standard texts on physical optics. Such a diffraction pattern, obtained by Larsson in 1929[1], of x-rays transmitted through a fine slit is shown in figure 2.

The impetus behind the research aimed at establishing a close analogy between x-rays and visible light declined rapidly after 1912, because in that year Max von Laue and his colleagues discovered that crystals act as three-dimensional diffraction gratings for x-rays.

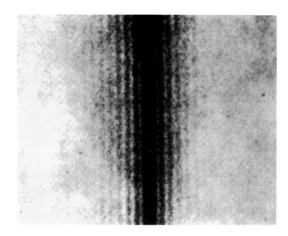

Figure 2. Diffraction pattern of a 5.5 μm wide slit obtained with 0.08 nm wavelength x-rays.

Not only did their work demonstrate unequivocally the electromagnetic nature of x-rays and that their wavelengths are of the order of 0.1 nm but it also laid the foundations of x-ray crystallography and crystal spectroscopy.

Production of X-Rays

Right from the early days of x-rays radiography has been the most widely employed x-ray technique. There are, however, many other applications of x-rays, and a knowledge of the different ways in which x-rays can be generated helps us to understand the underlying principles of these applications.

X-rays are usually produced by using high voltage to accelerate electrons emitted by an electron gun to bombard a metal target, as shown in figure 3. On striking the target, the electrons are rapidly decelerated and can interact with the target in two distinct ways to give rise either to a continuous or to a characteristic x-ray spectrum.

Continuous or 'white' radiation or bremsstrahlung is produced by electrons that are slowed down in passing through the strong electric fields near the atomic nuclei of the target material. Some electrons may approach a nucleus closely enough that all their kinetic energy is converted into x-rays at once. Others will lose their

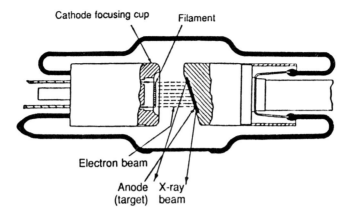

Cathode focusing cup   Filament

Electron beam

Anode   X-ray
(target)   beam

Figure 3. A conventional vacuum x-ray tube. Electrons emitted from a heated filament are accelerated towards the anode which is held at a high voltage.

energy more slowly through many encounters so that there is a more gradual production of x-rays. In the case where all the energy is converted in one encounter, the x-rays produced have the highest energy (i.e., shortest wavelength), while electrons losing their energy more gradually will give rise to a continuous spectrum of longer wavelengths. The range of wavelengths emitted in bremsstrahlung therefore depends only on the value of the accelerating voltage and not on the target material.

The characteristic x-ray spectrum differs from the continuous spectrum in that the wavelengths emitted are specific to the target material. The origin of the characteristic radiation is explained in terms of the electronic structure of the atom, which may be thought of as a small positively charged nucleus surrounded by a planetary system of electrons. Their orbits, 'shells', or more correctly their energy levels are named K, L, M, N, O and P in order of increasing distance from the nucleus. It is these energy levels that are characteristic of the atom and hence of the element. Emission of, say, the characteristic K radiation of an element requires that the energy of the bombarding electron is sufficient to remove an electron from the K level. When the atom has thus been ionised, an electron can transfer from higher energy levels to occupy the vacant K level and this transition results in the emission of radiation characteristic of the

K spectrum. Characteristic and continuous radiation are emitted together owing to the random energy distribution and random behaviour of the bombarding electrons.

An alternative method of exciting the characteristic spectrum is by irradiating the target with x-rays of energy sufficient to remove an electron from one of the energy levels. The x-rays so generated are called fluorescent x-rays. This radiation is of high spectral purity and contains relatively little continuous background, unlike radiation excited by electron bombardment.

X-rays may be specified either in terms of their wavelength or their energy: the choice depends on their application. Crystallographers, for example, are concerned with the dimensional aspects of x-rays and therefore specify x-rays in terms of their wavelength in nanometres (or traditionally, in Ångstrom Units (Å), where $10\,\text{Å}=1\,\text{nm}$). Many other practitioners, for example astronomers and radiographers, find it more appropriate to specify x-rays in terms of their energy in electron volts (eV). The relationship for conversion from one unit to the other is (wavelength in nm)=1.24/(energy in keV). The x-ray spectral region is loosely defined as spanning the wavelength range 0.01–300 nm. The shorter wavelengths overlap the γ-ray region and the longer wavelengths overlap the extreme ultraviolet region.

Applications of X-Rays

Radiography is the first and the foremost of the many applications of x-rays that have been developed during the past hundred years. It is also the most familiar one, with applications ranging from medicine to the detection of structural flaws in materials and to security checks on luggage. X-rays of wavelengths shorter than 1 nm (equivalent to about 1 keV) can penetrate submillimetre thicknesses of material, while in medical radiography x-rays with energies in the range 30–90 keV are commonly employed in order to allow imaging of the internal parts of the body. In industrial radiography energies in the MeV range are used for the inspection of steel castings of the order of a metre thick.

The intensity of the beam transmitted through the object under examination is dependent on the thickness and absorbing power of the material and on the wavelength of the x-rays used; in general, the higher the atomic number of the material, the greater is the absorbing power. This differential absorption is the basis of radiography and of other forms of x-ray imaging. In dentistry, for example, an x-ray beam that encounters a metal filling will be absorbed more than the beam that traverses the sound part of the tooth, and a shadow image of the filling will be registered on the film placed behind the tooth. In an early form of x-ray microscopy, but one which is still widely employed, the shadow image on the film or other x-ray sensitive material is enlarged optically or in an electron microscope to yield a highly magnified image. X-rays of wavelengths longer than 1 nm may also be used in radiography and x-ray microscopy but, because their penetrating power is so low, observations can only be made on very thin specimens.

There were two strands that led to Max von Laue's discovery of x-ray diffraction. It had long been presumed that a crystalline solid was composed of a three-dimensional lattice of atoms. Laue had calculated from the known number of molecules per unit volume that the average distance between atoms in a solid was between 0.1 nm and 1 nm. From Sommerfeld's calculations, previously re-ferred to, it was deduced that the wavelength of x-rays was in the range 0.01–0.1 nm. By analogy to the conditions where light is diffracted by a grating, Laue speculated that such a periodic atomic array should act as a three-dimensional diffraction grating for x-rays. In 1912 he was able to demonstrate that x-rays were indeed dif-fracted by crystals, and he also calculated how x-rays would be diffracted by a three-dimensional array. However, it was the father and son team of William Henry and William Lawrence Bragg who developed a simple relationship, known as the Bragg law ($\lambda = 2d\sin\theta$), between the spacing d of the crystal lattice planes giving rise to a diffracted beam, θ the angle of the diffracted beam and λ the x-ray wavelength. This equation enabled them to interpret the diffraction pattern and deduce the crystal structure, from the angular measurement of the diffracted beam, in terms of the ratio d/λ only. Alternatively, from a knowledge of the form of the crystal structure, the molecular weight, Avogadro's number and the density, it was a simple matter to calculate the spacings of the crystal lattice

planes. The x-ray wavelength could then be calculated by subs-
tituting the value of d in the Bragg law. The first wavelength to be
measured in this way was that of palladium K characteristic radiation, with a value of 0.0586 nm. This work laid the foundations both
of x-ray crystallography and of x-ray spectroscopy.

X-ray crystallography has provided us with the key to the
understanding and elucidation of the atomic and molecular structures of materials. Simple cubic structures such as that of sodium
chloride were discovered in the pioneering work by the Braggs.
Later, more complex structures were determined, such as the DNA
double helix. The determination of this structure was based on the
work of Maurice Wilkins and Rosalind Franklin at King's College
London.

Subsequent to Laue's discovery and the work of the Braggs,
investigations continued, albeit on a reduced scale, of the physical,
and particularly the optical, properties of x-rays. It was not until the
early 1920s that Arthur Compton at the University of Chicago
demonstrated that x-rays of wavelength 0.13 nm could be reflected
from polished surfaces, but only at small grazing angles of incidence
of several minutes of arc. This experiment heralded the birth of x-ray
optics, and was a demonstration of the phenomenon of the total
external reflection of x-rays. This occurs as a consequence of the fact
that refractive indices of materials for x-rays are slightly less than
one. It is analogous to the more familiar phenomenon of the total
internal reflection of visible light, for which materials have refractive indices greater than one. This work, together with the noteworthy development of ruled grazing incidence diffraction gratings for
x-rays, by Manne Siegbahn of Uppsala University, quickly led to the
direct measurement of x-ray wavelengths. By the late 1920s,
Joyce Bearden at Johns Hopkins University and others had developed the technique of precision wavelength measurement with ruled
gratings to such an extent that these methods yielded the most
reliable values of x-ray wavelengths.

We have seen that materials will emit radiation that is characteristic of their elemental composition when bombarded by electrons
or x-rays. Measurement of the intensity of the emitted characteristic
radiation by wavelength sensitive detectors is the basis of widely

employed methods for elemental analysis. In the 1950s and 1960s the UK played a leading part in the development of electron probe x-ray microanalysis (EPMA), particularly through Peter Duncumb and James Long at Cambridge University and Tom Mulvey at Associated Electrical Industries. In this technique a finely focused beam of electrons is used to bombard a volume as small as a cubic micrometre of a specimen. EPMA is a powerful method for identifying small impurities or particulates and for mapping the distribution of different elements in materials.

X-rays are used in fluorescent x-ray analysis to excite the characteristic radiation. This is a very sensitive method of analysis because the radiation is spectrally very pure. In the fluorescent analysis technique, those incident x-rays that have just sufficient energy to ionise the target material by removing an electron will themselves be highly absorbed by the specimen material. The spectrum of x-rays transmitted through a specimen will therefore show characteristic drops in intensity corresponding to the high absorption at particular energy levels. This is the basis of x-ray absorption spectroscopy.

The general effect of x-rays on living cells is a lethal one and is the basis of their therapeutic applications. The early workers in x-rays suffered from dermatitis and other diseases because they were not aware of the chemical and biological effects of x-rays. It is the ionising power of x-rays that is responsible for their chemical action, which may be utilised in a variety of applications.

The lithographic application of x-rays depends on their chemical modification of materials called photoresists, or resists. The interaction of the beam with the resist material induces structural changes, such as polymerisation, that alter the solubility of the irradiated areas. Although the underlying principles of x-ray and optical lithography are similar, the x-ray method has the advantage that much sharper images can be formed in the resist material than with light, because the resolution of a light beam is severely limited both by scatter in the resist and by diffraction. Consequently x-ray lithography is being increasingly employed to develop integrated circuits with finer structural features, a wide range of microstructures and other miniature components. A $0.2\,\mu m$ SRAM gate made by exposing x-rays through a mask fabricated by electron beam machining[2] is shown in figure 4. The resolution is limited by the gap between mask and wafer. A resolution of better than 40 nm is achievable for gaps less than 10 μm, but typical working distances today are 20–30 μm.

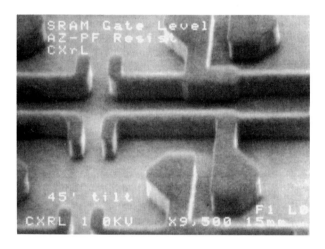

Figure 4. Scanning electron microscope image of a 0.2 μm SRAM gate made by x-ray lithography.

X-rays have also had a major impact in metrology, particularly in the evaluation of the fundamental atomic constants and latterly in precise dimensional measurements.

The most conspicuous and well defined feature of the continuous x-ray spectrum generated in an x-ray tube is the short wavelength or high energy limit. The shortest wavelengths are produced by those electrons whose energies are entirely converted into x-rays in single encounters with the nuclei of the target material. The short wavelength limit λ_0 is inversely proportional to the x-ray tube voltage V, and the latter is therefore directly proportional to the corresponding limiting frequency ν_0 of the x-rays. The energy of the incoming electron is eV, where e is the charge on the electron, so that $eV = h\nu_0 = hc/\lambda_0$, where c is the speed of light. The constant of proportionality h is Planck's constant. In quantum terms, $h\nu_0$ is the maximum quantum energy of the emitted x-ray. The ratio h/e can thus be determined from measurements of V and λ_0.

The Braggs used the value of the Avogadro constant to calculate x-ray wavelengths. Later, the inverse calculation was employed to yield a more precise value of the Avogadro constant when diffraction gratings became available that enabled x-ray wavelengths to be measured directly and very precisely. Knowledge of the precise value of the Avogadro constant provides a possible path that may ultimately lead to defining the kilogram in terms of an atomic constant, i.e., as an invariable number of atoms of a particular element. The kilogram is the only base unit in the International System of Units (SI) that is still defined in terms of a material artefact—a cylindrical block of platinum-iridium that fluctuates unpredictably in mass.

Today an even more direct method, which circumvents the need to measure the x-ray wavelength, is used to obtain lattice spacings. In 1965 Ulrich Bonse and Michael Hart developed an x-ray interferometer. This closely resembles an optical interferometer in which displacement of one component of the interferometer results in the movement of optical fringes. A displacement of half the wavelength of light (say 0.3 μm) results in the movement of a fringe by a distance equal to the fringe spacing. The number of fringes passing a detector is thus a measure of the displacement of the interferometer on a scale based on the wavelength of light. In the x-ray interferometer, a silicon crystal is employed to produce the interference pattern and displacement of one component of the interferometer by a distance equal to the lattice spacing of silicon (approximately 0.3 nm) will result in the movement of an x-ray fringe through one fringe spacing. A large displacement can be measured directly by optical interferometry, and by counting the x-ray fringes produced during the displacement the lattice spacing of silicon can be determined directly. The x-ray fringes provide a measurement scale a thousand times finer than can be obtained by optical interferometry, and x-ray interferometers can thus be used to measure displacements with accuracies of a few picometres.

The early successes of x-ray crystallography and the power of the technique appealed so widely that other aspects of x-ray physics were relatively neglected for many years. Several factors contributed to the post Second World War renaissance of x-ray physics.

In 1946 a large number of captured German V-2 rockets became available for scientific research. Solar x-ray emissions were detected above the Earth's atmosphere in 1948 with photographic plates mounted on these rockets. Ground based observation was not possible because of absorption of x-rays by the atmosphere. This was the start of x-ray astronomy. A prominent role in the early work was played by Herbert Friedman of the US Naval Research Laboratory in Washington. Ken Pounds at Leicester University and Robert Boyd at the Mullard Space Science Laboratory were UK pioneers of x-ray astronomy. Over the years spectacular advances in x-ray astronomy have revolutionised our understanding of cosmology.

Mechanisms whereby extraterrestrial x-rays are generated include bremsstrahlung resulting from acceleration of electrons in magnetic or electric fields. X-rays are also produced by the interaction of relativistic electrons or high energy cosmic-ray particles with starlight or with microwaves of the ubiquitous cosmic background that is believed to represent the remnant of the primeval fireball from which the universe expanded. A low energy photon, starlight or microwave, may in a single collision acquire sufficient energy to convert it into x-ray in the keV range. Thermal sources provide another type of extraterrestrial radiation of both characteristic and continuous x-rays. In many stellar sources, high energy electrons excite the moderately heavy atoms of a gas, causing the emission of characteristic lines. If the source is very hot, as is the case with neutron stars having surface temperatures of ten million degrees Celsius, all the electrons will be stripped from the nucleus resulting in the emission of continuous radiation.

Processes that occur in earthbound experiments in high temperature plasma physics are closely related to some of those that occur in the Sun and the stars. An important objective of one aspect of this research is the generation of thermonuclear energy by laser induced fusion. In this work extremely powerful laser beams are focused onto small glass spheres containing thermonuclear fuel. The action of the laser is to compress and thus heat the sphere to such an extent that the matter within mimics the processes that occur in stellar nuclear reactions. The x-rays that are generated provide a means of diagnosing the various processes that are taking place in this reaction. Another objective of plasma physics research is to generate x-rays for their own sake, such as in x-ray lasers. These lasers, which produce coherent beams of x-rays, are finding practical applications in the microscopy of biological materials, as well as in fundamental studies in plasma physics. The Lawrence Livermore National Laboratory in California and the Central Laser Facility of the Rutherford Appleton Laboratory in the UK, under the direction of Michael Key, are world leaders in this area of work.

The Daresbury Synchrotron Radiation Source (SRS), which started to operate in the early 1980s, was the first synchrotron in the world dedicated specifically as a source of x-rays and longer wavelength radiation. Earlier synchrotrons were mainly employed as particle accelerators for high energy physics research, although synchrotron radiation was first observed as early as 1947. In a synchrotron radiation source, a beam of electrons is accelerated to close to the speed of light within an evacuated doughnut shaped ring and is maintained in a circular orbit by means of external magnets. The circular motion of the electrons produces radiation which is emitted tangentially to the orbit and extracted through portholes in the ring. Synchrotrons can provide ultraviolet radiation and x-rays over 100 000 times more intense than from conventional x-ray tubes. Since it is in the form of a continuous spectrum researchers can select wavelengths most appropriate for their work. Because of these unique properties the rapid worldwide growth of interest in this research tool is not surprising.

The availability of the very high intensity continuum spectrum of synchrotron radiation has provided the opportunity to undertake research which was not possible or practicable by any other means. The few cases that are given here provide only a superficial insight into the wide range of applications of synchrotron radiation. For example, short exposure times make possible the study of transient phenomena such as the molecular changes that occur in muscular contractions. Another application of very great practical importance is the acquisition of an understanding of the mode of action of catalysts. These substances may sometimes be present in small quantities in a reaction vessel, but they can profoundly affect the rate at which chemical reactions take place. Studies with synchrotron radiation of subtle changes in the x-ray absorption spectrum have provided a unique insight into catalytic action. Other diverse applications of synchrotron radiation include x-ray microscopy and x-ray lithography, the latter for the manufacture of semiconductor devices and miniature mechanical components.

X-Ray Optics

The quest for an x-ray microscope provided the initial stimulus for the postwar resurgence of x-ray optics. X-ray transmission microscopy has much to offer in comparison with light microscopy: in principle up to a thousand times finer resolution, the capability of examining specimens that are opaque to light and a means of analysing the specimen material by differential absorption of x-rays. The two factors that rule out lenses, as used in optical microscopes, are that the refractive indices of materials are so close to unity that x-rays are barely deviated when passing through them, and that the greater the refraction the higher is the absorption in the lens.

Imaging by using total external reflection at grazing incidence from concave surfaces had been unsuccessfully attempted in the 1920s, and the idea was revived independently by Werner Ehrenberg at Birkbeck College and by Paul Kirkpatrick and Albert Baez at Stanford University in the late 1940s. The latter used two spherically concave mirrors to focus the x-ray beam as shown in figure 5. For specimens such as grids, which exhibit very high contrast, images demonstrating resolutions of a few micrometres were achieved. The idea was ahead of its time in that the technology for producing surfaces of adequate form and quality had not yet been developed. Nevertheless, in later years Kirkpatrick-Baez optics became widely employed.

An alternative approach to x-ray microscopy was pursued at the Cavendish Laboratory by Ellis Cosslett and Bill Nixon. Their expertise in the development of electron optics for electron microscopy led to the development of submicrometre sized x-ray sources, using electron lenses to focus an electron beam. The x-ray source was placed close to the specimen and the magnified image was obtained

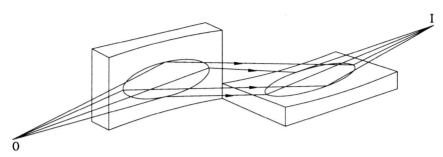

Figure 5. Kirkpatrick-Baez optics. X-rays diverging from a source O are focused to form an enlarged image at I after successive reflection from crossed concave cylindrical or spherical surfaces.

by projection of the radiograph onto a distant plane. By this means useful magnifications of 3000 were achieved.

Since the early 1960s there has been a continuing interest in the use of zone plates as a means of forming x-ray images, as an alternative to employing grazing incidence optics. The Fresnel zone plate is essentially a circular diffraction grating and consists of an array of concentric circles with radii proportional to the square roots of the consecutive integers 1, 2, 3.... The areas between the concentric circles are made alternately opaque and transparent to the incident radiation, as shown in figure 6, just as in the zone plates employed in visible optics. The development of the latter can be traced back to Lord Rayleigh in the 1870s, and the zone plate principle is described in the standard optics textbooks. The manufacture of x-ray zone plates has required the development of sophisticated techniques, for example similar to those used in the micro-

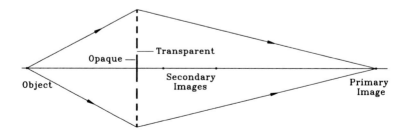

Figure 6. Imaging with a zone plate.

electronics industry. Several teams around the world, including one at King's College, have been in the forefront of this research and their work has led to the development of powerful x-ray microscopes capable of resolving features as small as 30 nm.

Although x-ray microscopy played a special role in the postwar revival of x-ray optics, there was also a growing demand for x-ray optical components for use in other applications. X-ray astronomers wanted x-ray telescopes of increasingly higher optical aperture and resolving power. Synchrotron users required optical components to focus and collect x-rays more efficiently. X-ray spectroscopists required gratings with high diffraction efficiencies for measuring the wavelengths of the characteristic radiations of the low atomic number elements as a means of analysing these materials by spectro-scopic methods, for analysing radiations emitted in plasma physics and for the spectroscopy of extraterrestrial radiations.

The demand for high quality x-ray optical components could not be met by the then current optical manufacturing capabilities: some components required surface form accurate to nanometres and surface finish to subnanometres. Not only did these manufacturing tolerance requirements fall outside the experience of the optical worker, but the measuring instruments for checking tolerances to these limits were not generally available. Parallel developments in optical manufacturing technology and measurement instrumentation by Albert Franks and his team at the National Physical Laboratory culminated in the development both of x-ray gratings with diffrac-tion efficiencies of 20%, an order of magnitude greater than previ-ously achieved, and other x-ray reflectors, including the first x-ray telescope with resolution less than an arc second[3] (figure 7).

This telescope consists of hyperbolic and parabolic figures of revolution, commonly employed in x-ray telescope optics, in a configuration devised by Hans Wolter in 1952, and illustrated in figure 8. It can be seen from the figure that the collection efficiency of the telescope is small because the x-rays are incident at a grazing angle. Radiation entering the central aperture and blocked by the baffle is wasted as it plays no part in the imaging process. To improve the collection efficiency, telescopes have been constructed in which the central aperture is filled with a series of confocal

Figure 7. Nickel-coated beryllium x-ray telescope of large internal diameter, figured and polished at the National Physical Laboratory. This was the first x-ray telescope having sub-arc second resolution.

concentric telescope shells; these are known as nested telescopes. The XMM (X-Ray Multi-Mirror) telescope is an example of such a telescope. It consists of 58 nested telescope shells within a 700 mm diameter outer shell, and collects x-rays incident at grazing angles over a range of 18 to 40 arc minutes.

A more fundamental way of improving x-ray collection efficiency depends on increasing the grazing angle of incidence at which x-rays will be reflected with high efficiency. This has been achieved by coating the mirror surface with a periodic structure consisting of thin layers of reflecting material, usually of high atomic number such as tungsten, separated by a low absorbing, low atomic number element such as carbon. The following example illustrates the principle. A polished tungsten mirror reflects about 1% of carbon K radiation (4.4 nm) incident at a grazing angle of 20°, falling to about 0.01% at normal incidence. Since the intensity is proportional to the square of the amplitude, an intensity reflectivity of 0.01% represents an amplitude reflectivity of 1%. Neglecting absorption, it would thus appear that 100 single surface reflections, if combined in

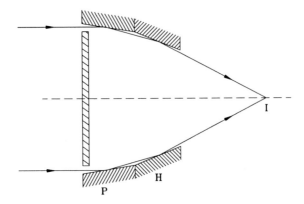

Figure 8. Ray path of a Wolter type I telescope consisting of parabolic (P) and hyperbolic (H) figures of revolution. The collection efficiency is small in the grazing incidence configuration.

the correct phase, would enable 100% reflection efficiency to be achieved at normal incidence. In practice, there are losses that occur as a result of absorption, as well as imperfections in the films and in their spacings due to the deposition methods. The path difference of reflections from successive layers must equal one x-ray wavelength (recall the Bragg law), so that precise fabrication is very demanding. Although the principle has been known for many years and is employed analogously in visible optics, it was not until the 1970s that Eberhard Spiller at IBM and Troy Barbee at Stanford University developed the deposition technologies to make the first successful multilayer mirrors. It is now possible to employ normal incidence optics for wavelengths greater than about 4 nm and to increase grazing angles significantly for shorter wavelengths.

The multilayer technique has been widely used in the manufacture of telescopes, x-ray microscopes and synchrotron components. It should be noted that a multilayer mirror reflects efficiently in a small wavelength region only, whereas total external reflection is effective over a much wider range of wavelengths.

Summary
In conclusion, we can chart the changes in emphasis that have taken place in the field of x-rays during the past century. Radiographic

applications became established from the very earliest days and have maintained their dominant position ever since. Research in x-ray optics and physics flourished until x-ray crystallography became a central activity subsequent to 1912. From then until after the Second World War, x-ray optics and spectroscopy were relatively low key activities. The postwar years witnessed a renaissance in x-ray optics fuelled by the demands of x-ray microscopy, plasma physics, x-ray spectroscopy, x-ray astronomy and synchrotron applications. We enter the second century since the discovery of x-rays confident in the knowledge that x-rays will play as important a role in the future as they have done in the past.

Bibliography

O Glasser "Wilhelm Conrad Röntgen" London: John Bale Sons & Danielson (1933)

A H Compton and S K Allison "X-Rays in Theory and Experiment" New York: D van Nostrand (1960)

V E Cosslett and W C Nixon "X-Ray Microscopy" Cambridge: University Press (1960)

H Winnick and S Doniach (eds) "Synchrotron Radiation Research" New York: Plenum Press 1980

A G Michette "Optical Systems for Soft X-Rays" New York: Plenum Press (1986)

M Elvis (ed) "Imaging X-Ray Astronomy: A Decade of Einstein Observatory Achievement" Cambridge: University Press (1990)

P B Kenway *et al* (eds) "Proceedings of the International Congress on X-Ray Optics and Microanalysis" *Inst. Phys. Conf. Series* **130** (1992)

"Annual Reports of the Central Laser Facility" Rutherford Appleton Laboratory

References

(1) A Larsson *Uppsala Univ. Årsskrift* **1** 97 (1929)

(2) P N Dunn "X-Rays Future—a Cloudy Picture" *Solid State Technology* **37**(6) 49 (1994)

(3) A Franks "X-Ray Optics" *Science Progress* **64** 371 (1977)

The Early History of X-Rays in Medicine for Therapy and Diagnosis

Richard Mould
Scientific Consultant

R esearchers at King's College London have long been associated with x-rays and thus it is most appropriate that the College should have hosted a Celebratory Conference exactly 100 years to the day after the momentous discovery by Wilhelm Conrad Röntgen (figure 1) on the 8th of November 1895 in Würzburg.

Herbert Jackson, then a demonstrator in chemistry at King's and later to become a professor, was responsible for one of the first two major improvements in the design of x-ray tubes. The first was by Alan Campbell-Swinton, an electrical engineer, who suggested the introduction of a metal anticathode to replace the glass end of the pear shaped vacuum tube as a target. Campbell-Swinton was also the first in the United Kingdom to produce a radiograph, on the 8th of January 1896. However, it was Jackson who proposed the use of a concave cathode to concentrate the electrons onto the target, thus designing the 'focus tube'. As to its value, James Gardiner in 1909, who was to be a future President of the Röntgen Society (1915–16), wrote in his review *The Origin, History and Development of the X-Ray Tube*, that Jackson's contribution "...marked the greatest advance that has been made since Röntgen's discovery 13 years previously...".

The following text and illustrations refer to only a very few examples of the achievements of the early years which led to today's advances in x-ray imaging and x-ray therapy.

X-RAYS: The First Hundred Years
Edited by Alan Michette and Sławka Pfauntsch © 1996 John Wiley & Sons Ltd

Figure 1. Röntgen photographed by Hanfstaengl of Frankfurt-on-Main and used as the frontispiece of an American textbook on x-rays of 1896. The author made the mistake of claiming that Röntgen was born in Holland rather than in Germany! Indeed the first newspaper (Vienna Presse) announcement called him Professor Routgen and stated that the discovery occurred in Vienna. However, this is not too surprising as many misconceptions followed the great discovery, not least that x-rays were the Philosopher's Stone and could transmute base metals into gold!

The Discovery of X-Rays

Röntgen's discovery of x-rays in the Physical Institute of the University of Würzburg occurred when he made an unexpected observation while experimenting with various Lenard and Crookes tubes. Some barium platinocyanide fluorescent material smeared on thin cardboard and lying some distance from one of the excited tubes, which was covered with black light tight paper, glowed visibly.

It did not take Röntgen long to discover that not only black paper but also other objects such as a wooden plank, a thick book and metal sheets, were penetrated by these rays. More important, however, he found, according to his biographer Otto Glasser, that

"Strangest of all, while flesh was very transparent, bones were fairly opaque, and interposing his hand between the source of the rays and his bit of luminescent cardboard, he saw the bones of his living hand in silhouette upon the screen. The great discovery was made".

On 1 January 1896 Röntgen wrote to scientific colleagues in several countries enclosing some example radiographs (figures 2 and 3) together with a reprint of his first communication *Eine Neue Art von Strahlen* which was published in December 1895 in the Sitzungsberichte der Physikalisch-Medizinischen Gesellschaft zu Würzburg and was set out in 17 numbered paragraphs. His second communication (1896) was a continuation of the first with additional paragraphs 18–21. His third and final paper (1897) was entitled *Further Observations on the Properties of X-Rays*.

Röntgen never published again on x-rays. He gave only one public lecture describing his discovery, in Würzburg on January 1896, where he gave a demonstration of a radiograph of the hand of

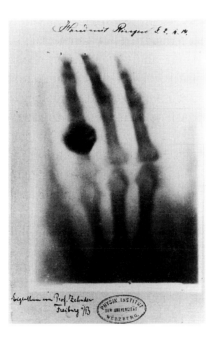

Figure 2. A radiograph of the hand of Frau Röntgen taken in 1895. This photograph was sent to Professor Ludwig Zehnder of Freiburg in Breisgau, who had been a former student of Röntgen.

Figure 3. An x-ray picture of a compass referred to by Röntgen in paragraph 14 of his 1895 paper. 'I have observed, and in some cases recorded photographically, many shadow pictures of this kind, the production of which occasionally affords a very special fascination. Thus for instance I possess photographs of the shadow of...a compass in which the magnetic needle is completely enclosed in metal...'

the famous anatomist Albert von Kölliker, who then proposed the term 'Röntgen rays' and called for three cheers for Röntgen. The audience cheered again and again. Many honours followed his discovery, most notably, in 1901, the first ever Nobel Prize for Physics. He is also commemorated on many coins and stamps (figure 4), and by medals and eponymous lectures named after him.

Figure 4. Röntgen commemorated on a 1939 stamp bearing the slogan "Fight cancer, cancer is curable".

Public Reaction

Public reaction was immediate and widespread following the announcement which circulated the world in January 1896 with headline captions such as: *Electrical Photography Through Solid Bodies* (Electrical Engineer, New York), *Illuminated Tissue* (New York Medical Record) and *Searchlight of Photography* (The Lancet).

However, not all the responses were favourable and the London Pall Mall Gazette stated "We are sick of the Röntgen rays…you can see other people's bones with the naked eye, and also see through eight inches of solid wood. On the revolting indecency of this there is no need to dwell".

Many lectures demonstrations were given during the months of 1896, often with a fee, donated to the speaker's favourite charity, charged to members of the audience who volunteered to have their hands, purses, etc. x-rayed. A typical advertisement for 'the wondrous rays' is seen in figure 5. This was distributed at an 1896 exhibition at the Crystal Palace, London.

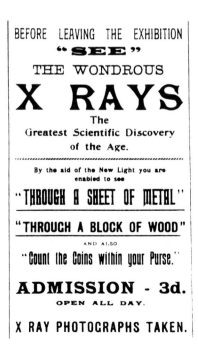

Figure 5. An 1896 advertisement

Radiography

A pear shaped x-ray tube of the type used by Röntgen is clearly seen in figure 6. This is the world's first published photograph, in a radiological journal *Archives of Clinical Skiagraphy*, of a radiographic setup. In the same issue of these *Archives* was the very good quality radiograph shown in figure 7. It was entitled "Child-skiagram of skeleton of full grown child aged three months. Note that the intestines, heart and liver cast definite shadows". The term 'skiagram' was used in 1896 for what we now call a radiograph.

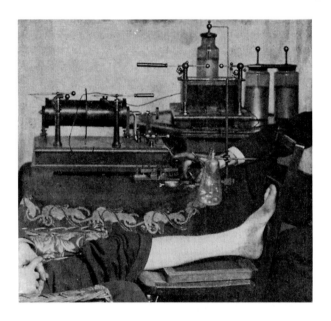

Figure 6. Radiography of the leg, from the 1896 *Archives of Clinical Skiagraphy*.

Test objects were used from the very early days for both radiography and fluoroscopy. These included the hand of the physician or technician, often leading to fatal effects with skin cancer following the formation of radiation ulcers (figure 8); amputation treatment was not generally successful in arresting disease. Skeleton hands and forearms in frames with soft tissue simulated by silver paper were an improvement but even when these devices, sometimes called Osteoscopes or Chiroscopes (the forerunners of today's physics test phantoms), were used the x-ray tube was often unshielded.

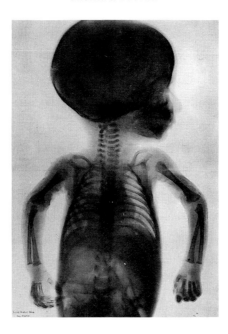

Figure 7. An 1896 skiagram of a three month old baby; 14 minutes exposure.

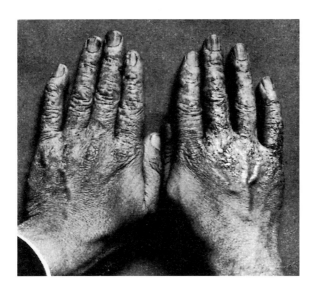

Figure 8. The hands of the x-ray martyr, the physician Mihran Kassabian of Philadelphia. He described his hands as 'showing the result of chronic x-ray dermatitis' and stated that he began work in 1899 and that the photograph was taken in 1903. He died in 1907.

Other test objects which were popular in the early years were small animals of which figures 9 and 10 are two examples. The frog was the most frequent of these for illustrating image quality but it is recorded that a snake (reported to be chloroformed!) was used in an experiment to investigate the optimum length of the spark gap of the x-ray apparatus.

Figure 9. An 1896 radiograph of a frog taken in San Francisco. (above)

Figure 10. An 1896 radiograph of a rattlesnake. (right)

Radiography and fluoroscopy were not limited to the environment of the hospital or the physician's private consulting rooms but were also practiced on military campaigns (figure 11) and in non-medical applications. The latter included in 1896 the detection of false gems and of the contents of suspicious packages (for what was termed 'infernal machines', i.e., terrorist bombs) in customs halls.

Figure 11. Radiography in the Sudan campaign in the desert some 1200 miles from Cairo (1898). The x-ray tube was described as being suspended by means of an ingenious holder. The use of the inverse square law is to the advantage of the operator on the right but not of the soldier near the patient! It was at this installation that the casualties from the battle of Omdurman were radiographed.

The x-ray tube in figure 11 is an unshielded focus tube of the type seen in the 1896 advertisement of figure 12. The Herbert Jackson concave cathode developed at King's College is clearly seen in this advertisement and also in figure 13, which shows a later design of an x-ray tube with a heavy anode.

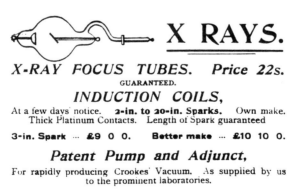

Figure 12. An 1896 advertisement.

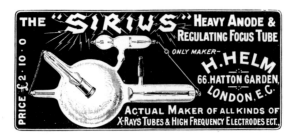

Figure 13. An 1913 advertisement.

Herbert Jackson of King's College

Jackson, later Sir Herbert Jackson FRS and President of the Röntgen Society (1901–03), has already been mentioned with regard to his focus tube and it is of interest to record statements in 1896 on his work at King's College. The photographer Snowden Ward who published the first x-ray textbook in England (entitled *Practical Radiography*) states in his chapter *A Brief History:* "The tubes for producing x-rays were the subject of much experiment…In February the 'focus' tube was suggested for this purpose by Herbert Jackson and on March 4th its marvellous capabilities were shown at the Society of Arts. The great advantages of this tube caused it to replace, at once, all other patterns".

Jackson's work on x-rays was not limited to tube design. In the first issue (May 1896) of the *Archives of Clinical Skiagraphy* the editor Sydney Rowland, who was also Special Commissioner to the *British Medical Journal* for investigation of the applications of the new photography to medicine and surgery, discussed in his introduction probably the world's first fluoroscope (then called cryptoscope), described by Enrico Salvioni of Perugia. He made the following remarks concerning the choice of a fluorescent material, having stated that neither Salvioni nor Thomas Edison had achieved optimal results: "Great has been the success of observers in other countries, it has remained for England to produce the instrument in its perfected and simplified form, and for Mr Herbert Jackson, of King's College, to fix on the particular salt which gave the best results—results besides which all previous attempts are cast into the shade. The salt he employs is the platinocyanide of potassium".

Fluoroscopy

An early 1896 construction of a fluoroscope is shown in figure 14 in conjunction with the type of pear shaped x-ray tube used by Röntgen in 1895. This geometrical shape of the instrument was retained for many years and, for example, features in the 1919 United States Army X-Ray Manual.

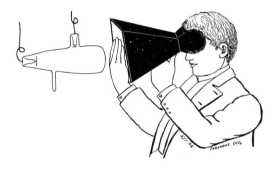

Figure 14. An 1896 fluoroscope.

Such fluoroscopes were used by Thomas Edison and figure 15 shows the public waiting to see a demonstration at Edison's 'Beneficent X-Ray Exhibit' at the 1896 New York Electrical Exposition of the Electric Light Association. In 1896 Edison also described an 'x-ray lamp' in which the inner surface of the lamp was covered with a fluorescent material fused on the glass. When this lamp, which was in effect an x-ray tube, was connected to an induction coil Edison stated that one could obtain "light production at the rate of 0.3 watt per candle power". However, the results of manufacturing such lamps were disastrous and Edison soon noticed that his assistant Clarence Dally suffered from loss of hair and from skin ulcers and he therefore abandoned work on his x-ray lamp. This was, however, too late for Dally, who became the first of the x-ray martyrs, dying in 1904 at the age of 39.

An important piece of equipment for fluoroscopy is the patient table and the first tilting table is shown in figure 16, dating from 1898 in Paris. Prior to this, patients were draped over chairs and ordinary tables in order to position them so that a successful x-ray examination (radiographic or fluoroscopic) could take place.

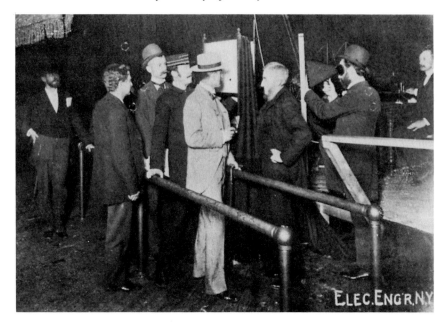

Figure 15. Thomas Edison's 'Beneficent X-Ray Exhibit' 1896.

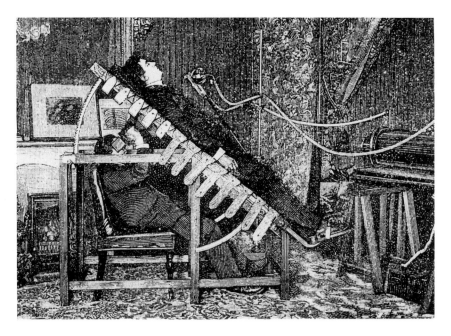

Figure 16. Tilting table for fluoroscopy, Paris 1896.

Photofluoroscopy, Cine-Radiography and Arteriography

The first photofluoroscope was designed in 1896 by Julius Mount Bleyer of Naples and the use of the camera for cine-radiography was also demonstrated that year by John Macintyre of Glasgow. This involved the sensitive film passing underneath the aperture in a case of thick lead covering the cinematograph. This opening corresponded to the size of the picture, and was covered with a piece of black paper upon which the limb of an animal, such as a frog, could be photographed. In this way, Macintyre showed the Glasgow Philosophical Society 40 feet of film demonstrating the movement of a frog's leg.

Arteriography using cadavers and injections of radio opaque material such as mercury formed the basis of experiments as early as 1896. Figure 17 shows a later example (1904) from the *Archives of the Roentgen Ray* by Alfred Fryatt of Melbourne, Australia, illustrating one of a series of stereoscopic arteriogram pairs which included the hand and the heart.

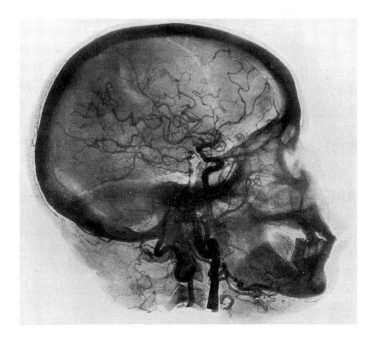

Figure 17. An 1904 arteriogram.

Radiation Protection

The necessity for radiation protection was not immediately recognised and even when it was by some x-ray practitioners many ignored the warnings documented in the literature. The dramatic effect of working with unprotected hands has already been shown in figure 8; many other similar cases have been described. However, as early as 1898 the Röntgen Society of London formed a Committee on X-Ray Injuries and in America in 1903 a Protection Committee was proposed within the American Roentgen Ray Society.

Figure 18. Lead protective wear, London 1910.

By 1910 many x-ray operators were wearing lead protective clothing such as in figure 18 and by about this time the use of x-ray shielding, sometimes only partial as in figure 19, had become more widespread. However, it was not until 1921 and 1922 respectively that national committees in Britain and the USA adopted protection rules and recommendations. Figure 20 is from this era and shows full shielding of the x-ray tube and collimation of the beam.

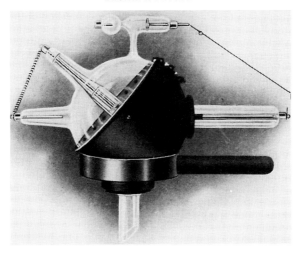

Figure 19. Partially shielded x-ray tube made by Müller of Hamburg in 1915.

Figure 20. An advertisement from an issue of 1921 *Journal of the Röntgen Society.*

X-Ray Therapy

It was also during the early years that many patients with non-malignant conditions were treated with x-rays. These included ringworm (figure 21), lupus, hæmangioma, syphilis and tuberculosis. In addition, x-rays were also used for cosmetic purposes to remove unwanted hair. The long term effects of such treatments are now well documented and, for example, a seven year old girl treated for ringworm in 1913 was diagnosed with basal cell carcinoma of the scalp in 1937 which was treated but recurred in 1952.

IMPROVED SHIELD

for the treatment of

RINGWORM,

by SABOURAUD'S METHOD.

(Designed by Dr. A. SAVILL.)

Price of Shield, with Arm and Universal Joint for fixing to any upright stand, complete with three stops and Wire Spreader £4 6/-
Ball Jointed Tube Holder—shown..   ..   13/6
Suitable Stand for carrying the Apparatus, 30/-
Box of Sabouraud's Pastilles 9/-

Figure 21. An 1906 advertisement in the *Journal of the Röntgen Society.*

The same x-ray apparatus used for diagnosis was originally also used for therapy but, as technology improved so that higher x-ray energies could be obtained and longer lasting targets designed, special therapy tubes became available. Figure 22 shows a typical apparatus of 1903 which, some 20 years later, had changed out of all recognition to become what was known as an x-ray 'cannon' (figure 23). This was possible because of the development in 1913 by William Coolidge of the hot cathode x-ray tube, which immediately made the gas tube obsolete for all applications: for diagnosis and therapy in medicine and also for industrial purposes.

Figure 22. X-ray therapy department at the London Hospital.

Figure 23. An American x-ray cannon of the 1920s manufactured by the Wappler Electric Company of New York. The cylinder was lined with 0.25" lead and all openings were lead flanged. The treatment table had a blower to constantly circulate air, carried out from the table by an exit pipe, in the tube chamber.

Radiation Units

No chapter on the early days of x-rays in medicine would be complete without a mention of some of the more than 50 radiation measurement units proposed prior to the international acceptance in 1937 of the roentgen as a unit for both x-rays and gamma rays.

Quantitative measuring instrumentation was a great improvement on the initial qualitative methods of visual inspection of the colour of the x-ray tube when it was operational, self-exposure of the hand or forearm to measure a skin erythema, or the use of a skeleton phantom as previously mentioned.

Chemical colour change following irradiation formed the basis of several radiation units. For example the pastille of 1904, which was used well into the 1930s, was based on the use of a small capsule of barium platinocyanide, purchased in small booklets called radiometers (figure 24). There were two standard tints, A for unexposed and B for the standard epilation dose.

The H-unit of 1902 used a fused mixture of potassium chloride and sodium carbonate. The Holzknecht radiometer was equipped with a scale and it was stated that the epilation dose was 1.25 B

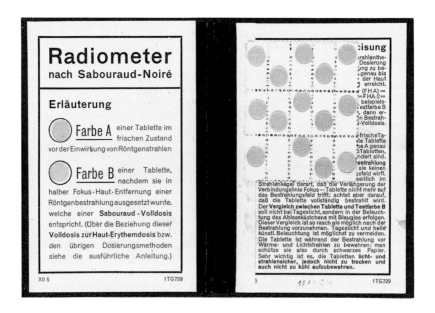

Figure 24. Booklet of pastilles.

which equalled 6 H and that the erythema dose was 2.5 B which equalled 12 H. However, in practice these units were varied in use by individual physicians and a 1911 dosimetry survey of '13 radiologists of repute' was entitled *Idiosyncrasy and Dosage* to emphasise the then current problems.

Photographic film blackening was another radiation effect upon which a dose unit was based: the X-unit of 1905. Small strips of film, known as Kienböck strips (figure 25), were exposed on the patient's skin and the density of the developed film was compared to an arbitrary scale of blackening using an instrument called a quantimeter. The conversions given in 1932 for x-rays filtered by 0.5 mm of copper were 65 X = 2.5 B = 12 H = 550 x-ray roentgens.

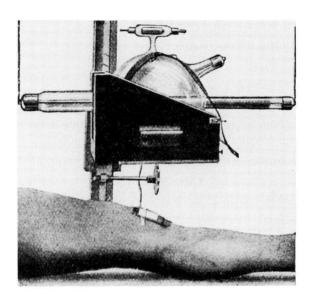

Figure 25. Partially shielded x-ray tube of 1918. A Kienböck strip can be seen just below the knee. This method of dosimetry was suggested in 1905 by Robert Kienböck of Vienna. Earlier, in 1902, Williams Rollins of Boston had suggested silver bromide film, for measurement of dose, as a radiation protection standard.

Other radiation effects upon which units of dose were based included ionisation, fluorescence, temperature variation, x-ray current, heating (energy deposition per unit volume), and the change in electrical resistance of a layer of selenium.

X-Ray Quality

X-ray quality was a term used to describe the penetrative power. The original proposal (1900) was only qualitative in terms of hard (highly penetrating), medium and soft. These terms were used for many years, but by 1905 several 'penetrameters' had been designed such as the Benoist (1902) radiochromator which consisted of a thin disc of silver surrounded by 12 aluminium steps of increasing thickness. When this instrument was placed behind a fluorescent screen the luminosity of the central silver disc was compared with the steps of the aluminium ladder. Soft x-rays were steps 2 or 3 and hard x-rays were steps 7 or 8. This and similar devices fell into disuse when the concept of the half value layer was enunciated in 1912, allowing a much more quantitative description of x-ray quality. The half value thickness is the thickness of a material which reduces the initial intensity of a beam of radiation by 50%.

Summary

This chapter has described the discovery of x-rays and early experiments concerned with their properties and applications. Although Röntgen himself determined many of the important properties of x-rays, much of the work described was carried out before the nature of x-rays was completely understood. Other applications described later in this book required a more thorough understanding of x-rays and their properties which has been achieved by a century of research.

Bibliography

More detailed reviews, including illustrations, may be found in the following publications.

R Eisenberg "Radiology an Illustrated History" St. Louis: Mosby (1992)

O Glasser "Wilhelm Conrad Röntgen and the Early History of the Roentgen Rays" Berlin: Julius Springer (1933)

O Glasser (ed) "The Science of Radiology" London: Baillière Tindall & Cox (1933)

E R N Grigg "The Trail of the Invisible Light" Springfield: Charles C Thomas (1965)

R F Mould "A History of X-Rays and Radium" Sutton: IPC Business Press (1980)

R F Mould "A Century of X-Rays and Radioactivity in Medicine with Emphasis on the Photographic Records of the Early Years" Bristol: Institute of Physics Publishing (1993)

R F Mould "The Early Years of Radiotherapy with Emphasis on X-Ray and Radium Apparatus" *British J. Radiology* **68** 567 (1995)

R F Mould "The Early History of X-Ray Diagnosis with Emphasis on the Contribution of Physics 1895–1915" *Physics in Medicine & Biology* **40** p 1741 (1995)

X-Ray Microscopy

William Nixon
Cambridge University

A s we celebrate the 100th anniversary of Röntgen's discovery of x-rays, their use in medical applications and many other scientific investigations has become ubiquitous. For example, the short wavelengths of x-rays make them suitable for microscopy when there is a need for viewing the interior of the specimen at high magnification. Several forms of x-ray microscopy have been used with varying degrees of success and are described in this contribution. Projection x-ray microscopy, including early work at the Cavendish Laboratory, is highlighted along with more recent work using zone plate microscopes in conjunction with modern x-ray sources.

Contact Microradiography

Röntgen and other early x-ray scientists were not able to deflect x-ray beams with mirrors or lenses. As a result, the first form of x-ray microscopy to be suggested was contact microradiography[1]. In this method, which does not require x-ray focusing, the specimen is placed in contact with a detector which records the internal detail of the specimen when it is illuminated with a beam of x-rays. In the early days photographic plates were used but they have now been replaced by photoresists, which are capable of much better spatial resolutions. After development the plate or resist contains an x-ray transmission map of the specimen which may then be viewed in an optical microscope, an electron microscope or, recently, an atomic force microscope. The magnification is all done with the microscope used for viewing, so that the image resolution can be no better than that of the microscope employed. When an electron or an atomic force microscope is used the limiting resolution, typically several

X-RAYS: The First Hundred Years
Edited by Alan Michette and Sławka Pfauntsch   © 1996 John Wiley & Sons Ltd

tens of nanometres, is caused by diffraction due to the finite thickness of the specimen.

Contact microradiography has been used with success in several medical and other applications, most notably in recent years by Tony Stead and Tom Ford of Royal Holloway College who have used the technique for studying the internal structures of a variety of plant cells[2,3].

Reflection X-Ray Microscopy

In the early 1920s Arthur Compton showed that it was possible to reflect x-rays, so long as they were incident at a very small angle to a highly polished surface. However, the use of these so called glancing angles leads to very distorted images. For two-dimensional focusing the distortions can be reduced by using two glancing incidence mirrors, but not entirely eliminated as the mirrors must follow one after the other so that they give different magnifications.

Despite these problems early work on reflection x-ray microscopy was carried out over 40 years ago by Paul Kirkpatrick and Alfred Baez at Stanford University[4]. However, the x-ray optics did not produce better results than other methods of microscopy, and further research was dropped. More recently, x-ray astronomy has used similar optics, carried in satellites, for observations above the earth's atmosphere. The development of x-ray telescopes has been very successful, leading to a national award in the United States to Kirkpatrick and Baez for their pioneering work in x-ray optics.

Projection X-Ray Microscopy

The projection method of microscopy was amongst the many techniques suggested in the 1930s, but no experimental work was done until after the Second World War. The first detailed exploration of this topic was carried out by myself as a research student from 1949 to 1952 in the Cavendish Laboratory at Cambridge University[5]. The results were very promising and have led on to a variety of techniques. The original equipment I used during this research is on permanent display in the Museum of the Cavendish Laboratory and an exact copy is also on permanent display in the Science Museum in London. Many visitors come to see this equipment and some of them must ask "How did you get such beautiful results from that heap of junk?" The results do have a beauty in their own right, and any experimental, as opposed to commercially available, equipment usually looks like junk.

A schematic diagram of the equipment for projection x-ray microscopy is shown in figure 1. The physical form of the equipment is shown on the left with the electron gun at the bottom, followed by the condenser lens and the objective lens. These two magnetic lenses demagnify the electron source from some 50 μm at the electron gun to an image of 0.1 μm at the target. It is the interaction of the electrons with the target which forms the x-ray source. The vacuum connection is shown on the left and at the same place on the right of the column there is an observation window for viewing a fluorescent screen which aids mechanical alignment of the two magnetic lenses. A one foot scale is also shown (in those days, the metric system had not taken over in science in the UK). Magnetic lenses have much lower spherical aberration than electrostatic lenses and, therefore, can produce an electron current density a thousand times higher than that available from electrostatic lenses for the same focal spot

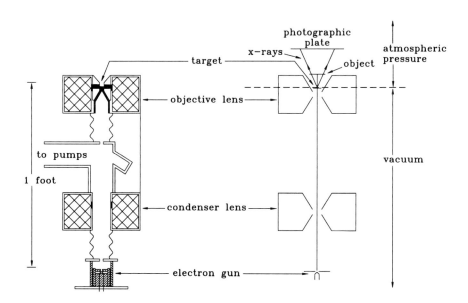

Figure 1. A schematic diagram of an x-ray projection microscope.

diameter. This means that exposure times using magnetic lenses are up to a thousand times shorter than with electrostatic ones.

The schematic diagram on the right of figure 1 shows how the thin foil x-ray target separates the vacuum region, necessary for the electron beam, from the atmospheric pressure region containing the specimen and the photographic plate. The divergence of the x-rays from the point source gives a magnification equal to the distance to the photographic plate divided by the distance to the object, both measured from the x-ray source.

Two images with a slight displacement of the specimen can be combined to show a stereographic effect with the internal detail seen in three dimensions. In addition there is a very large field of view with this method of microscopy, which cannot be seen in a publication with normal page sizes, but which may be appreciated on displays or when projected in lectures.

Results from many areas of investigation have been obtained with projection x-ray microscopes. Two examples are shown here. In

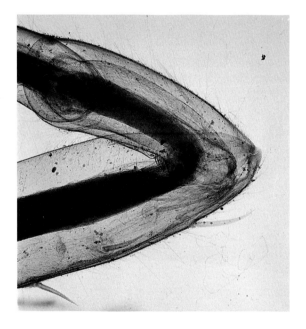

Figure 2. An x-ray projection image of a spider's knee.

figure 2 a magnified x-ray image of a spider's knee shows the internal detail available all in focus at once due to the large depth of field of this technique.

A higher magnification image is shown in figure 3, where the specimen is bull spermatozoa, osmium fixed and gold shadowed. This image was obtained with an exposure time of 5 min and an electron accelerating voltage of 7 kV. The sperm tail is approximately 0.3 μm in diameter. Some detail may be seen in the region between the head of the sperm and the full length of the tail.

Figure 3. An x-ray projection image of bull spermatozoa. The size of the whole micrograph is 75 μm by 50 μm.

These results were published in 1955, forty years ago. Since then, there have been several developments in projection x-ray microscopy and related methods. From 1954 to 1957 Peter Duncumb, also working in the Cavendish Laboratory, combined the magnetic electron optics techniques of the x-ray projection microscope with the scanning electron microscope principle developed by Charles Oatley and Denis McMullen at Cambridge University. He also incorporated x-ray detectors, as used by Raymond Castaing in Paris in an analytical instrument employing electrostatic electron lenses and an optical microscope for specimen viewing but with no electron scanning.

The research at the Cavendish showed the desirability of making an instrument for metallurgical and other applications. This was done by Duncumb and David Melford at the Tube Investments Research Laboratory and the commercial scanning x-ray microanalyser was produced by the Cambridge Instrument Company under the name "Microscan"; almost 100 instruments were made. The research was also incorporated in the development of the scanning electron microscope by the Cambridge Instrument Company with Gary Stewart moving from the University to industry to be in charge of this development. The first scanning electron microscope was marketed in 1965 under the name "Stereoscan" and has continued as a successful product, including x-ray analytical facilities, over the thirty years to 1995.

Recent developments in projection x-ray microscopy have been pioneered by Jacques Cazaux and his colleagues[6] at the University of Reims, France. They have added a cooled charge coupled device (CCD) two-dimensional x-ray detector to a scanning electron microscope, so that the x-ray projection principle can yield analytical results. The speed of the CCD detector allows information to be extracted very quickly, in contrast to the much older photographic methods. This modern form of projection x-ray microscopy allows the imaging of materials of interest to the petroleum industry, the semiconductor industry, geologists, metallurgists and biologists. The instrument also allows quantitative elemental mapping of specimens by comparing the images obtained using x-rays of two different energies lying on either side of an absorption edge of the element of interest.

X-Ray Microscopy using Zone Plates

X-ray microscopy, as a tool for the high resolution imaging of biological and materials science specimens, began to receive increased attention in the early 1980s following the development of high intensity synchrotron radiation sources, which is described later in this volume. This revival was aided by technological advances in the manufacture of optics and by the realisation that lengthy and complicated specimen preparation, usually needed in electron microscopy, can lead to artefacts in the image.

Specimen preparation in electron microscopy is necessary for several reasons. Since they are charged, electrons scatter readily in matter and hence can penetrate only small distances into solids (about 100 nm for electrons with energies of about 100 keV) so that specimens have to be sectioned (thinned). Some attempts at high voltage electron microscopy have been made to alleviate this, with varying degrees of success[7]. Electron attenuation lengths are very similar for elements close to one another in the periodic table so that heavy metal stains have to be used to give contrast in an image, especially of biological material. To prevent loss of the electron beam, due to scattering between the source and the specimen, electron microscopes must operate under high vacuum which means that specimens have to be dehydrated. Environmental containment chambers have not been very successful at overcoming this problem although cryoelectron microscopy has[8], but this is a difficult technique to do correctly. All this means that it is not usually possible, in electron microscopy, to image specimens in anything like their natural environment, and it is this problem which x-ray microscopy is primarily designed to overcome.

Image contrast in x-ray microscopy is possible for unstained specimens since absorption lengths can change rapidly from element to element. In addition, in the wavelength range between the oxygen and carbon K absorption edges at 2.3 nm and 4.4 nm, respectively, water is much less absorbing than material containing carbon. Absorption lengths in this wavelength range, known as the water window, are typically about 1 μm and so thick hydrated specimens can be imaged. These specimens can be at atmospheric pressure since the x-ray absorption length in air is a few millimetres.

For visible light solids (and liquids) have refractive indices upwards of about 1.3, whereas gases have refractive indices of effectively unity. A beam of light in air incident on a glass surface (refractive index about 1.5) at an angle of 45° is bent through an angle of about 17° on entering the glass. This bending, or refraction, occurs because of the large difference in the refractive indices of the two materials and is why lenses, such as those in spectacles, can be used to focus beams of light. At x-ray wavelengths, however, the refractive indices of all materials (gases, liquids and solids) are very close to unity and differences between them are very small. A beam of x-rays, with wavelength 3.5 nm, entering the same glass block at the same angle would be bent through an angle of only 0.18°, so that hardly any focusing of x-rays would take place with a lens made of glass. Incidentally, at most x-ray wavelengths refractive indices are slightly less than unity, and concave rather than convex lenses would have to be used to cause a beam of x-rays to converge. Absorption in the lens material would also mean that hardly any of the x-ray photons would actually be transmitted by the lens. Thus conventional refractive lenses are not suitable for x-rays.

Microlenses have been suggested[9,10] but the small sizes, a few tens of micrometres in diameter and about a micrometre thick (necessary to give reasonable x-ray transmission), lead to focal lengths of a few centimetres—very long compared to the lens diameter and resulting in resolution capabilities no better than several micrometres.

Mirrors are also commonly used as optical components at visible wavelengths. However, at normal incidence (as used in everyday mirrors) x-ray reflectivities of all materials are very small and typically less than one in a hundred thousand x-ray photons will be reflected. Thus it is not possible to use normal mirrors. High reflectivities can be obtained at grazing incidence angles but these give very severely aberrated images unless complicated systems are used. One such aberration, familiar to many people who have to wear spectacles, is astigmatism which causes a point in the object to be turned into a line in the image. High reflectivities can also be achieved by coating a surface with successive layers of high and low atomic number materials, resulting in a multilayer mirror[11]. Just like naturally occurring crystals these work by Bragg reflection, but can be used at longer wavelengths. The performances of multilayer mirrors are limited by the ability to deposit smooth layers of the correct thicknesses. Smoothness is particularly important if high reflectivity is to be maintained, as a roughness of just 5% of the layer thickness can decrease the normal incidence reflectivity by more than a factor of six. For wavelengths shorter than a few nanometres each layer must be only two or three atoms thick, which means that a single atom out of place can cause a roughness of up to 50%. This makes the manufacture of short wavelength multilayer mirrors very difficult.

Because of the difficulties involved with making and using x-ray optics employing refraction or reflection, most modern high resolution x-ray microscopes use optics based on the phenomenon of diffraction. This occurs when waves are obstructed by small objects, of the order of the wavelength of the radiation, causing the disturbance to spread beyond the limits of the geometrical shadow of the object. A diffraction grating consists of an array of equally spaced parallel lines and spaces; each line causes a beam of light to be diffracted. In certain directions, the diffracted beams will add together, while in other directions they will cancel one another. The angles at which the resulting intensity maxima occur depend on the wavelength of the incoming radiation—the angle is smaller for a shorter wavelength. Thus, if a diffraction grating is placed between a monochromatic light source and a screen, a series of light and dark lines will be seen on the screen. If the lines of the grating are made closer together the diffraction angle becomes larger.

Now imagine that a grating consists not of straight lines but of a series of concentric circles. Radiation incident on opposite edges of this circular grating will be diffracted through equal but opposite angles, so that beams will cross on the axis. If, in addition, the lines get closer together further from the centre then the diffraction angle increases outwards and it can be arranged so that all the diffracted beams cross the axis at the same point—the radiation is brought to a focus. This circular diffraction grating is known as a zone plate (figure 4) and the radius of an individual zone is proportional to the square root of the zone number, counting outwards from the centre.

Because the diffraction angles depend on wavelength the focal length of a zone plate is inversely proportional to the wavelength of the radiation. The zone plate should therefore be illuminated with monochromatic radiation, but even then it does not have a single focus. This is because combinations of 3, 5... zones acting together (third, fifth... diffraction orders) cause foci at distances 3, 5... times smaller than the first order focus, which is when each zone acts individually. Even numbers of zones acting together do not result in foci as the combinations then cancel one another out, so long as the

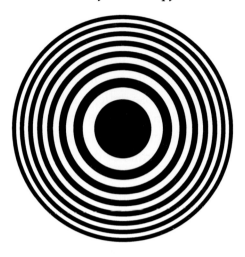

Figure 4. The structure of a zone plate.

zone boundaries are positioned correctly. Each focus contains only a fraction of the incident radiation; if alternate zones are totally absorbing and totally transmitting then only 10% goes into the first order. Zone plates are therefore inherently inefficient, although it is possible in principle to enhance the efficiency by careful control of the thickness and shape of the zones. Unfortunately, because of the small sizes involved, it is difficult to achieve the necessary control and the maximum experimental efficiency which has been achieved is about 15%[12].

Because of the multiplicity of foci and because some radiation passes straight through without being diffracted (zero order) it is usually necessary to use a zone plate with a central stop and an axial aperture (figure 5). The stop prevents undiffracted radiation from the centre of the zone plate from contaminating the image, and the aperture removes the zero order from the outer parts of the zone plate and the out of focus other orders.

Spatial resolution is defined as the minimum discernible distance between two points in an object. The spatial resolution which can be achieved with a zone plate is essentially equal to the width of the smallest, outermost, zone. Thus, to achieve high resolution, it is necessary to manufacture patterns with very small linewidths. In order to retain good imaging properties the zone boundaries have to

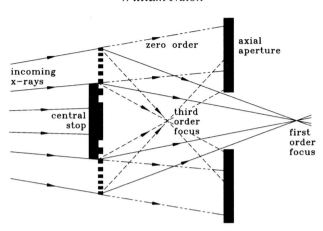

Figure 5. The use of a central stop and an aperture to remove the unwanted diffraction orders of a zone plate.

be placed in their correct positions to within a small fraction of the outermost zone width. With modern manufacturing techniques, which are derived from the lithographic methods for making micro-circuits, it is difficult to maintain the required accuracy over distances greater than about $150\,\mu m$. Current x-ray zone plates have maximum radii of about $75\,\mu m$ and minimum outermost zone widths of about $30\,nm$; this leads to focal lengths of a few hundred micrometres to about a millimetre in the water window.

Thus, using zone plates, spatial resolutions of about $30\,nm$ are now achievable in x-ray microscopy. The present aim is to reduce this to about $10\,nm$. Although this does not compare favourably with the resolution of about $0.5\,nm$ possible in electron microscopy, the less complicated specimen preparation means that in many cases x-ray microscopy may be preferable. X-ray microscopy should be seen as complementary to, rather than competing with, electron and optical microscopies.

Present day high resolution transmission x-ray microscopy uses a condenser optic (normally a zone plate but mirrors are sometimes used) to focus the x-rays onto a specimen and an objective zone plate to form an enlarged image on film or a CCD camera (figure 6). A disadvantage of this type of x-ray microscopy is that, since the objective zone plate is inefficient, only a few percent of the x-rays

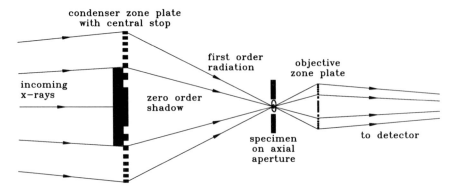

Figure 6. A schematic diagram of a transmission x-ray microscope using zone plates.

transmitted by the specimen can be used to form the image. This increases the radiation dose to the specimen. However, the whole image is formed at once so that results can be obtained quickly; with synchrotron radiation images can be formed in a few seconds. This technique has been highly developed by Günter Schmahl and his colleagues at the University of Göttingen in Germany, who have obtained images of specimens in near natural enviroments, including chromosomes, macrophages, lipid membranes and soil colloids[13].

In scanning transmission x-ray microscopy radiation damage is lessened by removing the post specimen zone plate (figure 7). Now, the objective zone plate focuses the x-rays to a small probe across which the specimen is mechanically scanned or, in a recent development, the probe can be scanned across the specimen[14]. The transmitted x-rays are detected by, for example, a proportional counter and the image is built up, raster fashion, pixel (picture element) by pixel. Using synchrotron radiation it is possible to form each pixel in a few milliseconds. The focused x-ray probe can also be used to excite other processes (e.g., photoelectrons, fluorescent x-rays) which can give spatially resolved information about a specimen. Compared to transmission x-ray microscopy, however, this method is slow. Scanning transmission x-ray microscopy has been developed by teams from the State University of New York, led by Janos Kirz, and from King's College. These microscopes have been used to study a range of specimens including chromosomes, malaria

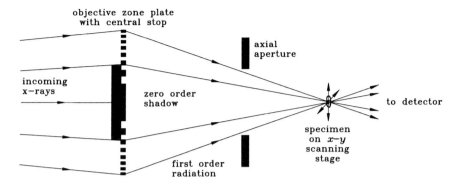

Figure 7. A schematic diagram of a scanning transmission x-ray microscope.

infected cells, sperm, muscle and, in materials science, cement, silica gels and zeolites[13]. Other microscopes are now being developed including several at the Advanced Light Source in Berkeley and the European Synchrotron Radiation Facility in Grenoble.

The process of mineralisation in structural and connective tissue is the subject of considerable worldwide investigation on both healthy and diseased systems. One aspect of this research is the imaging of mineralised tissue at medium to high spatial resolution. Scanning x-ray microscopy is well suited to this task, as it is possible to form quantitative elemental maps of specimens by comparing images taken at several different x-ray energies around absorption edges. Images of calcified tissue which map the distribution of calcium at a spatial resolution of about 50 nm can be obtained in just a few minutes. Pioneering work in this form of imaging is being carried out by Chris Buckley from King's College London[15], in collaboration with colleagues from the Institute of Orthopædics in London and the National Synchrotron Light Source at Brookhaven National Laboratory, Long Island. He is studying the early stages of diseases such as arthritis, tendonitis and osteoporosis (brittle bone disease).

Several biomineralised specimens have been examined in this way and a typical image is shown in figure 8. The specimen is a 0.1 μm thick section of human tendon taken from a patient suffering from tendonitis (calcification of tendon tissue). The causes of the onset of mineralisation of healthy tissue are not known, and the only

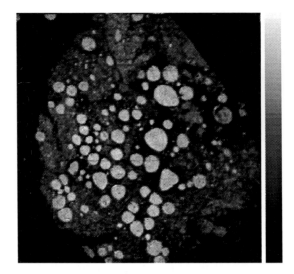

Figure 8. A quantitative map, obtained using scanning transmission x-ray microscopy, of the distribution of calcium in a 0.1 μm thick section of shoulder tendon from a patient suffering from tendonitis. The image size is 350×350 μm and the peak brightness corresponds to a calcium concentration of 5.5 μg cm^{-2}.

in active treatments for the condition are surgery or the administration of steroids. The mapping of the distribution of calcium in such specimens will provide information relevant to research on the biological reaction to pathogenic mineralisation of tissue. Other elements and compounds can be mapped in a similar fashion; for example DNA can be mapped by taking images around the carbon absorption edge.

Although most modern work on high resolution x-ray microscopy has been carried out using synchrotrons, these are very expensive sources which are national or international facilities. There is thus little chance of developing commercially viable x-ray microscopes based on such sources, but there is an increasing awareness of the necessity to develop suitable laboratory based microscopes. The most promising developments for transmission x-ray microscopy and scanning transmission x-ray microscopy are based on a plasma sources[16], and there is a good chance that one day in the not too distant future x-ray microscopes using such sources will become as common (perhaps) as electron microscopes.

Lensless X-Ray Microscopy

Another form of microscopical x-ray imaging in which major advances have been achieved recently is x-ray holography, first proposed by Baez in 1952. Unlike visible light holography, however, it has not yet been possible to obtain three-dimensional images because of the coherence properties of current x-ray sources. The principle advantage, at the moment, over other forms of high resolution x-ray microscopy is the ability to form images without having to focus the x-rays. The principal difficulty lies with reconstructing the holograms.

In its simplest form, Gabor, or in line, holography, the object beam is formed from the x-rays scattered by the specimen and the reference beam by x-rays transmitted by the specimen (figure 9). The recording surface (photoresist) is placed in the far field of the object. To date, only synchrotrons have been used to form high resolution holograms of real specimens, but x-ray lasers[17] and laser generated plasmas have been used for test specimens. The holograms can be reconstructed either by magnifying in an electron microscope and then using a visible light laser or by digitising the photoresist followed by computer reconstruction. The digitising can be done either on a magnified copy of the photoresist or directly with an atomic force microscope. The most advanced work on these techniques has been carried out by Denis Joyeux and Francois Polack[18] in France and, particularly, by Malcolm Howells and Chris Jacobsen[19] in the United States.

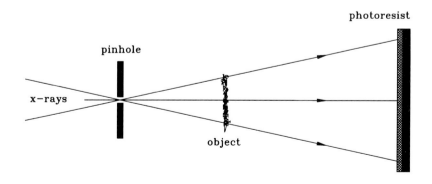

Figure 9. The arrangement for Gabor x-ray holography.

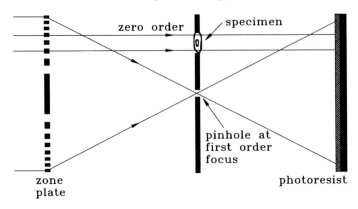

Figure 10. The arrangement for Fourier Transform x-ray holography.

Fourier transform x-ray holography (figure 10) has also been carried out with a synchrotron source[20]. In this, the specimen is placed in the first order focal plane of a zone plate such that the object beam is formed from the zero order radiation and the reference beam from the first order radiation.

Summary

X-ray microscopy, in its diverse forms, has advanced beyond all recognition since it was first suggested by P Goby in 1913. This progress has largely been due to advances in the technology of x-ray sources and optics. There is every reason to expect that x-ray microscopy will soon provide a routine analytical technique to complement and supplement other forms of microscopy, e.g., optical, electron and atomic force.

Acknowledgement

The sections from "X-Ray Microscopy using Zone Plates" onwards were written by Alan Michette of King's College London.

Bibliography

VE Cosslett & WC Nixon (eds) "X-Ray Microscopy" London: Cambridge University Press (1960)

G Schmahl & D Rudolph (eds) "X-Ray Microscopy" Berlin: Springer (1984)

PC Cheng & GJ Jan (eds) "X-Ray Microscopy: Instrumentation and Biological Applications" Berlin: Springer (1987)

AG Michette "X-Ray Microscopy" *Reports on Progress in Physics* **51** 1525 (1988)

D Sayre *et al* (eds) "X-Ray Microscopy II" Berlin: Springer (1988)

PJ Duke & AG Michette (eds) "Modern Microscopies: Techniques and Applications" New York: Plenum (1990)

MR Howells *et al* "X-Ray Microscopes" *Scientific American* **264**(2) 42 (1991)

AG Michette *et al* (eds) "X-Ray Microscopy III" Berlin: Springer (1992)

VV Aristov & AI Erko (eds) "X-Ray Microscopy IV" Moscow: Bogorodski Pechatnik (1995)

References

(1) P Goby "New Applications of Röntgen Rays: Microradiography" *C. R. Acad. Sci.* **156** 686 (1913)

(2) TW Ford *et al* "The Use of Soft X-Ray Microscopy to Study the Internal Ultrastructure of Living Cells and their Cellular Organelles" *X-Ray Microscopy IV* (AI Erko & VV Aristov, eds) Moscow: Bogorodski Pechatnik p276 (1995)

(3) AD Stead *et al* "The Use of Soft X-Ray Contact Microscopy Using Laser Plasmas to Study the Ultrastructure of Moss Protonemal Cells" *X-Ray Microscopy IV* (AI Erko & VV Aristov, eds) Moscow: Bogorodski Pechatnik p289 (1995)

(4) P Kirkpatrick & AV Baez "Formation of Optical Images by X-Rays" *J. Opt. Soc. Am.* **38** 766 (1948)

(5) WC Nixon "X-Ray Microscopy" *Contemporary Physics* **2** 183 (1961)

(6) J Cazaux *et al* "Progress in X-Ray Projection Microscopy" *Microscopy and Analysis* p19 (May 1994)

(7) JN Turner *et al* "Design and Operation of a Differentially Pumped Environmental Chamber for the HVEM" *Ultramicroscopy* **6** 267 (1981)

(8) M Stewart "Electron Microscopy of Biological Macromolecules: Frozen Hydrated Methods and Computer Image Processing" *Modern Microscopies: Techniques and Applications* (PJ Duke & AG Michette, eds) New York: Plenum Press p9 (1990)

(9) AG Michette "Optical Systems for Soft X-Rays" New York: Plenum Press: p29 (1986)

(10) S Suehiro *et al* "Refractive Lens for X-Ray Focus" *Nature* **352** 385 (1991)

(11) E Spiller "High Resolution Imaging with Multilayer X-Ray Mirrors" *X-Ray Microscopy IV* (AI Erko & VV Aristov, eds) Moscow: Bogorodski Pechatnik p544 (1995)

(12) EH Anderson & D Kern "Nanofabrication of Zone Plates for X-Ray Microscopy" *X-Ray Microscopy III* (AG Michette *et al*, eds) Berlin: Springer p75 (1992)

(13) J Kirz *et al* "Soft X-Ray Microscopes and their Biological Applications" *Quarterly Reviews of Biophysics* **28** 33 (1995)

(14) AG Michette *et al* "A Scanned Source X-Ray Microscope" *Meas. Sci. Technol.* **5** 555 (1994)

(15) CJ Buckley "The Measuring and Mapping of Calcium in Mineralised Tissues by Absorption Difference Imaging" *Rev. Sci. Instrum.* **66** 1318 (1995)

(16) AG Michette *et al* "Laser Plasma Sources for X-Ray Microscopy" *X-Ray Microscopy IV* (AI Erko & VV Aristov, eds) Moscow: Bogorodski Pechatnik p355 (1995)

(17) J Trebes *et al* "Demonstration of X-Ray Holography with an X-Ray Laser" *Science* **238** 517 (1987)

(18) D Joyeux *et al* "Principle of a 'Reconstruction Microscope' for High Resolution X-Ray Holography" *Proc. SPIE* **1140** 399 (1989)

(19) C Jacobsen *et al* "X-Ray Holography using Photoresists: High Resolution Lensless Imaging" *X-Ray Microscopy III* (AG Michette *et al*, eds) Berlin: Springer p244 (1992)

(20) I McNulty *et al* "First Results with a Fourier Transform Holographic Microscope" *X-Ray Microscopy III* (AG Michette *et al*, eds) Berlin: Springer p251 (1992)

X-Ray Microanalysis

James Long
Cambridge University

M icroanalysis, a word dating from the last century, was originally defined as the analysis of small quantities of material,
typically milligrams as opposed to the gram sized samples used for
conventional chemical analysis. The meaning of the term has
changed subtly over the years: broadly, it covers the identification
and quantification of small quantities of material, answering the
questions "what?" and "how much?" For example, the analysis of
minute dust particles is an important part of environmental research.
Many modern techniques are also able to provide an answer to the
question "where is it?", that is, of locating inclusions or giving
information on the distribution of a particular element or constituent
within a matrix. It also includes trace analysis in which we may be
examining a very minor constituent, either dispersed throughout a
bulk sample or perhaps located in small concentrated inclusions
within it.

X-Ray Spectra

That x-rays have played a major role in this field is directly due to the
ease with which they may be generated and to the regular nature of
their spectra which became clear in the classic work of Moseley just
before the First World War[1,2]. Prior to 1912, when Friedrich and
Knipping[3] and von Laue[4] demonstrated that x-rays were electromagnetic radiation, the 'quality' of x-rays was determined by measuring the absorption coefficient in aluminium. 'Hard' and 'soft'
x-rays had high and low penetrating power respectively. Barkla and
Sadler[5] had already shown that elements emitted 'characteristic'
radiation, each element giving rise to x-rays whose absorption

coefficient was lower than that of the radiation of the element immediately preceding it in the periodic table. Moseley's great contribution was to show that the wavelengths of the characteristic radiations decreased systematically through the periodic table and that atomic number was not merely an integer showing position in the table but a fundamental constant bearing a precise relationship to the characteristic wavelength and thus to the structure of the atom.

Moseley's experiments were made while he was a demonstrator in Sir William Bragg's department at Manchester using a form of the x-ray spectrometer which Bragg and his son, W L Bragg[6], had set up following von Laue's work. The apparatus is shown diagrammatically in figure 1. In the x-ray tube, target materials were carried

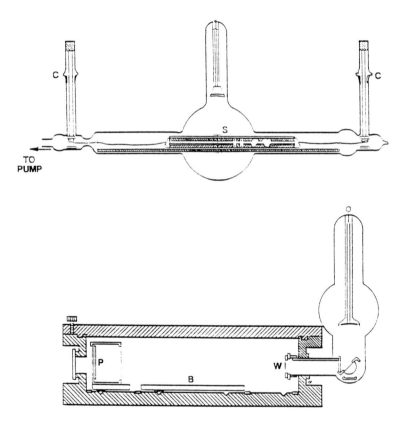

Figure 1. Two views of the apparatus used by Moseley to examine the characteristic spectra of the elements[2]. B: crystal table; C: greased cones; P: photographic plate; S: slit, W: window.

on a small trolley which could be pulled along by threads wound on bobbins turned by means of the greased cones. This form of apparatus had already been used by Kaye[7] in earlier studies of absorption coefficients.

X-rays generated by bombardment of one of the metal specimens passed through the slit and, after diffraction from a suitably oriented crystal (potassium ferrocyanide) mounted on the table, were recorded as as a series of lines on a photographic plate. The position of a line and the dimensions of the spectrometer allowed the angle through which the incident beam had been deviated at the crystal to be determined. The wavelength, λ, of the radiation producing the line was then given by the Bragg equation, $\lambda = 2d\sin\theta$, where θ is half the angle of deviation and d is the distance between diffracting layers in the crystal.

Figure 2 is a classic picture taken from Moseley's paper showing a series of exposures of the K spectra of elements from calcium to copper. It is also, incidentally, the first record of a chemical analysis with x-rays. The spectrum from brass shows the lines of both copper and zinc while the cobalt specimen is shown to be impure and to contain minor amounts of iron and nickel.

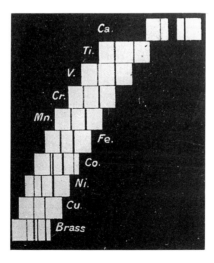

Figure 2. Photographic record of K spectra recorded by Moseley. Individual spectra are placed approximately in register. The brass shows nickel and copper lines; the cobalt contained both nickel and iron as impurities.

Figure 3 shows the wavelengths of the principal lines of the first three x-ray series discovered by Moseley, plotted against the atomic number. In addition to the wavelength, a second scale shows the photon energy in electron volts. Photon energy is often used as an alternative because it represents exactly the difference in the energy levels within an atom concerned in the production of a characteristic x-ray photon. If an electron whose binding energy is E_1 is ejected from an atom by electron or x-ray bombardment, and the resultant vacancy is filled by an electron dropping from a level of binding energy E_2, then the energy of a characteristic x-ray photon produced by the jump will be E_1-E_2. Photon energy and wavelength are related by $E=12390/\lambda$ where the wavelength is in Ångstroms and the energy is in electron volts. This relation applies to any electromagnetic radiation. Thus the characteristic line of copper has a wavelength of 1.54 Å and a photon energy of 8000 eV. The energy E_I required to ionise the atom and to create the vacancy is clearly greater than that of the characteristic line and is denoted by the dotted lines in figure 3.

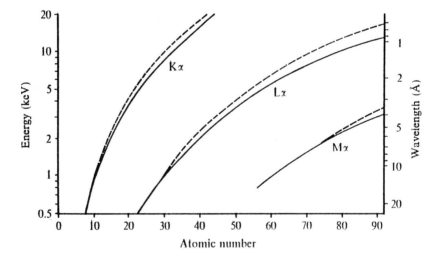

Figure 3. Wavelength and energy of the principal K, L and M series characteristic lines as a function of atomic number (full lines). The excitation energy (dotted lines) is always greater than the emission energy because it corresponds to the complete removal of of a K, L or M electron from the atom as opposed to a jump between levels in the atom. (Courtesy of Dr S J B Read).

Chemical Analysis with X-Rays

The basis of almost all microanalysis with x-rays is thus the identification of an element by the wavelength or photon energy of its characteristic emission and the determination of concentration in the sample by measurement of the intensity of that radiation. In general, it is very difficult to calculate concentration directly from intensity and practical quantitative analysis is always performed by comparing the intensity from the unknown sample with that from a standard in which the concentration of the element is known. Then, to a first approximation, the ratio of the concentrations in the specimen and standard is given by the ratio of the measured intensities. This relationship is only valid when the specimen and standard have similar matrix compositions. Where these differ it is necessary to make corrections for differences in absorption of the emergent x-rays and for other effects.

The intervention of the First World War, in which Moseley was killed, slowed the progress of x-ray spectroscopy, but in the 1920s and 30s it developed rapidly in the hands of a number of workers, notably de Hevesy, whose textbook *Chemical Analysis with X-Rays and its Applications*[8] gives a vivid picture of the achievements of these early years.

Figures 4 and 5 show components of a typical experimental setup of the time. The sample was either coated on the anode of the tube and bombarded directly with the electron beam or mounted very close to the anode[9] so that it was irradiated with intense primary radiation, emitting the characteristic spectrum as secondary or 'fluorescence' radiation. Although the characteristic spectra are identical in the two modes of excitation there is an important difference: electron bombardment produces not only characteristic x-rays from ionisation of inner atomic levels, but also a continuous background or 'bremsstrahlung', generated by slowing down of electrons in the target. When the atoms in the specimen are ionised by primary x-rays this radiation is absent. This has an important practical consequence: the lines in the spectrum from elements present in low concentrations tend to be obscured by the background with electron excitation whereas they stand out clearly and can be measured much more easily in the fluorescence radiation. The spectrometer (figure 4) used

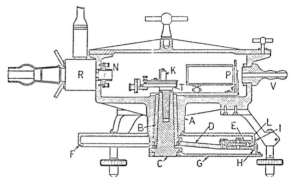

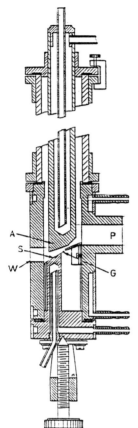

Figure 4. Siegbahn's vacuum crystal spectrometer. R: x-ray tube; N: entrance slit assembly; K: crystal; P: photographic plate (above).

Figure 5. The Alexander-Faessler x-ray tube. Electrons from the filament G are accelerated to the anode A. A specimen mounted on the water cooled surface S subtends a large solid angle at the anode, maximising the incident x-ray intensity. Fluorescence x-rays from the specimen emerge from the window W (left).

a large single crystal, usually of rocksalt, quartz, calcite or mica and the spectrum was recorded on photographic plates or film[10].

One of the most significant results to emerge was the discovery by Coster and de Hevesy[11] of the new element, hafnium, in minerals rich in zirconium. Figure 6 is a reproduction of the photographic recording of the L spectrum of hafnium obtained by these workers.

X-Ray Fluorescence Analysis of Bulk Samples

In addition to giving a low continuum background, the indirect excitation of x-ray spectra has the great practical advantage that the specimen does not need to be in the high vacuum of the x-ray tube; only a rough vacuum in the x-ray spectrometer is necessary to prevent significant absorption of the longer characteristic wave-

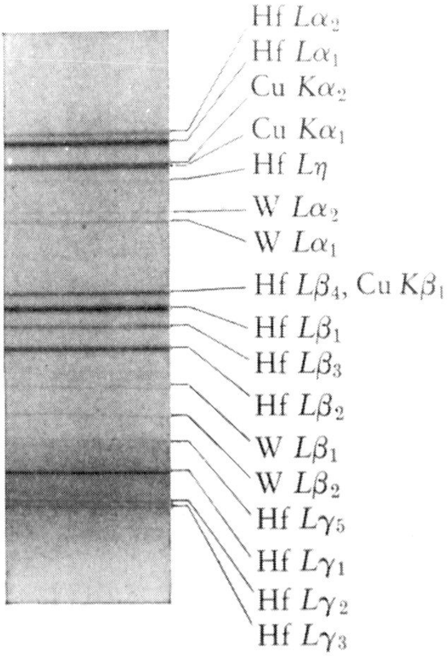
Hf $L\alpha_2$
Hf $L\alpha_1$
Cu $K\alpha_2$
Cu $K\alpha_1$
Hf $L\eta$
W $L\alpha_2$
W $L\alpha_1$
Hf $L\beta_4$, Cu $K\beta_1$
Hf $L\beta_1$
Hf $L\beta_3$
Hf $L\beta_2$
W $L\beta_1$
W $L\beta_2$
Hf $L\gamma_5$
Hf $L\gamma_1$
Hf $L\gamma_2$
Hf $L\gamma_3$

Figure 6. Photographic recording of the L spectrum of hafnium(8).

prevent significant absorption of the longer characteristic wavelengths. X-ray fluorescence (XRF) analysis, however, only became established as a routine analytical tool after the Second World War when x-ray detectors, particularly proportional and scintillation counters, and the associated electronics became available. The introduction of x-ray photon counting, initially with Geiger counters as in the instrument constructed by Friedman and Birks(12), transformed the quantitative measurement of x-ray intensities by eliminating the tedious procedures of development and microdensitometry of photographic emulsions. It was then possible to construct an apparatus which, once set up, could be operated by relatively unskilled personnel.

Figure 7 shows the basic arrangement of an XRF analyser equipped with a flat crystal Bragg spectrometer. The x-ray tube is typically operated at a power level of 1–3 kW while the anode material is ideally chosen to give strong characteristic radiation at an energy just greater than the the critical excitation energies of the elements to be determined. For example, a chromium tube is often

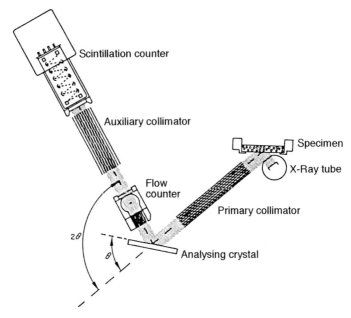

Figure 7. Schematic arrangement of an x-ray fluorescence spectrometer. The flow counter detects low energy x-rays but is equipped with an x-ray transparent rear window so that high energy x-rays can pass into the scintillation counter.

used for light elements such as calcium, potassium, sulphur and phosphorus. Alternatively, where a broad spectrum of energies is required for excitation, a target of high atomic number, e.g., gold or tungsten, may be used to maximise the intensity of the continuum.

The Philips Company of Eindhoven played a prominent part in the development of commercial XRF equipment. For many years, the PW 1520 instrument, introduced in 1956, and its derivatives (figure 8) were in widespread use in the ore mining and metallurgical industries. Each of these instruments was equipped with a single Bragg spectrometer with proportional and scintillation detectors for measurement of low and high energy photons respectively. The spectrometer was used with a helium atmosphere although later versions used a vacuum to reduce absorption. The manually operated specimen carrier allowed the sample and a standard to be interchanged. Spectra were recorded by scanning through a range of Bragg angles and recording the resultant spectrum on a chart re-

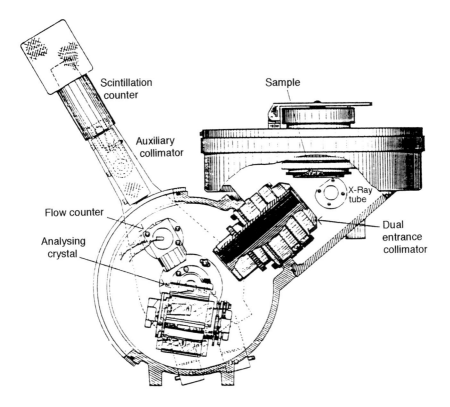

Figure 8. Cross section of the Philips PW 1410 x-ray fluorescence spectrometer equipped with interchangeable analysing crystals and collimators, flow proportional counter and scintillation counter.

corder. For more accurate quantitative work the spectrometer could be set to each Bragg peak in turn to alternately measure the standard and the sample.

As the stability of x-ray generators was improved, it became possible to measure an increasing number of samples between successive standardisations. A further instrumental development was the automation of specimen changing and spectrometer control, at first with hardwired controllers with teleprinter output of x-ray counts. In the instrument described by Hall[13], nine values of Bragg angle could be set on servo transmitters and selected in turn by a switching circuit. By 1963, the introduction of the Philips PW 1212 (figure 9), equipped with programmed selection of 24 mechanically

Figure 9. The Philips PW 1212 automatic x-ray spectrometer. This equipment is an excellent example of precomputer hardwired automation of the early 1960s. Parameters appropriate to recording each selected characteristic line, e.g., choice of analyser crystal, detector and pulse height analyser settings are preset on switches and selected sequentially. (Courtesy of Philips Ltd).

commercial instruments. With the advent of microcomputers in the late 1960s, software control of both data acqustion and processing became possible in XRF as in many other techniques. Today, such features are standard among the products of the major manufacturers, Philips and Siemens in Europe, Applied Research Laboratories (ARL) in the USA and Shimidzu and Rigaku in Japan. Philips (PW 2400), Siemens (SRS 3000) and ARL (ARL 9400), for example, all offer sample changers with a capacity of more than 100 samples at one loading, while angular reproducibility of the order of 0.0001 degree is achieved by using optical encoders.

A sample to be analysed in an XRF instrument may, in principle, be any material that can be presented to the x-ray beam. In practice, the form of the sample is constrained by the size of the area practice, the form of the sample is constrained by the size of the area viewed by the spectrometer and the need to control the geometry of the irradiated area in quantitative analysis. Typically, samples are prepared as discs approximately 3 cm in diameter, although for precious objects, e.g., archaeological artefacts, which must be examined nondestructively, the whole specimen is inserted into a suitably designed specimen chamber. Powdered samples are usually compacted with a hydraulic press into a disc, either with a small amount of an organic binder or with a backing of boric acid to increase strength. Liquid specimens may be examined using a sample cup with a thin x-ray transparent mylar base, through which both primary and secondary radations are readily transmitted. Clearly, it is not possible to use a vacuum in the spectrometer with most liquid samples and, in order to produce a low absorption path, the spectrometer is flushed with helium.

An important requirement for quantitative analysis is some means of correcting for matrix effects, i.e., for the influence of the bulk composition of the specimen on the intensity of the radiation from a given element contained in it. In the analysis of some materials, e.g., mild steel, the composition of the matrix is dominated by the major element, so that calibration curves for minor elements, once prepared, can be used to quantify concentrations in all such samples for a given instrument. The situation is very different with specimens such as natural rocks where there are generally several major elements present, e.g., oxygen, sodium, magnesium, aluminium, silicon, calcium and iron. The proportions of these elements vary between different rock types which results in significant differences in their x-ray absorption properties. The intensity per unit concentration for a given element is thus not the same for all rocks. This is particularly true for the light elements, whose characteristic radiation is of low energy and is thus strongly absorbed in the specimen. A further problem may arise in a disc made from a compressed powder because flaky minerals, such as mica, tend to concentrate at the surface which distorts the measured composition.

A widely used method of overcoming this problem in major element analysis, originally due to Claisse[14], is to fuse the powdered rock sample with 3–5 times its own weight of a borate flux and to pour the resultant melt into a mould so as to produce a thin glass disc. Variants of this method, including the addition of a heavy element such as barium or lanthanum[15], are used to minimise the differences between the x-ray absorption, and hence the matrix correction factors, of different specimens. Correction of the remaining small inter-element effects can then be achieved by solving a set of simultaneous equations in which the coefficients are determined either by calculation or semi-empirically. With the much greater computing power now available, it is possible to calculate matrix corrections over a wider range of compositions.

Simultaneous measurements of several elements means using a corresponding number of Bragg spectrometers, which is limited by cost and, ultimately, by the space around the specimen. Nevertheless, such a setup allows rapid analysis and measured intensity ratios are insensitive to variations in the primary x-ray beam intensity. These advantages stimulated the development of multiple spectrometer designs. The instrument developed by Adler and Axelrod[16] allowed simultaneous identification of four elements while ARL of California produced the VXQ design with 9 channels equipped with focusing curved crystal spectrometers. Commercial manufacturers now offer instruments with up to 28 separate channels which are well suited to repetitive analysis of a given group of elements, for example in process control. On the other hand, the readily adjustable single channel instrument, with software control of a fast servo operated spectrometer, generally offers greater versatility in research and in a wide range of analytical problems.

The development of energy sensitive solid-state detectors has opened up alternative possibilities for parallel recording. Such 'energy dispersive' spectrometers, while not possessing the excellent energy resolution of the Bragg spectrometer, have the advantages of high collection efficiency, absence of moving parts and, compared with multiple spectrometer instruments, considerably lower cost. Because of the high efficiency, only a low powered x-ray tube is required for excitation and many analysers have been constructed using radioisotopes as the primary source. Energy dispersive instruments are very easy to use, as illustrated by the spectra shown in figure 10. Here, separate measurements on two British 'copper' coins, each taking about 30 seconds, show clearly the change in composition made in the early 1990s. The 1989 coin is made from the traditional coinage bronze containing copper, tin and zinc. The 1994 coin, on the other hand, although showing a strong copper emission, has a beauty that is only skin deep. The low intensity iron K peak reveals the presence of an inner core of steel, more in keeping with the face value than the much more expensive bronze.

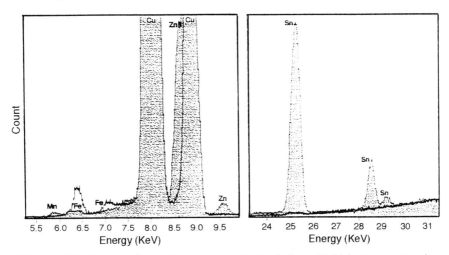

*Figure 10. Energy dispersive spectra obtained from British 'copper' coins (1989 shaded, 1990 unshaded). Note that the iron K radiation from the steel core of the 1994 coin is strongly attenuated by the copper plating. (Courtesy of Oxford Analytical Instruments).

Energy dispersive XRF analysis is now used quantitatively in many fields. Figure 11 illustrates its application to the characterisation of porcelain[17]. The genuine and rare Cheng Hua porcelain of the Ming Dynasty was recovered from the wreck of the Spanish galleon Concepcion, lost in 1641. It is clearly distinguished from faked copies by the characteristic ratio of rubidium to strontium determined by excitation with a 1 mCi source of cadmium-109.

A comparatively recent development with particular application to the analysis of thin layers and multilayers is total reflection XRF, in which a flat sample is illuminated by a well collimated beam of x-rays at glancing incidence, as illustrated in figure 12. This

X-Ray Microanalysis

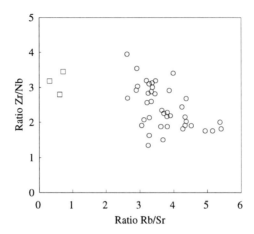

Figure 11. Ratio of the element intensities in antique Ming Dynasty porcelain (circles) and fake reproductions (squares)[17].

technique was first explored by Yoneda and Horiuchi[18] and later by Aiginger and Wobrauschek[19] who used an optically flat substrate on which a drop of aqueous solution, containing 10–100 ng of a dissolved metallic element, was evaporated to form a surface film. With an angle of incidence less than the critical angle (a few milliradians), at which the incident beam suffers total external reflection, absorption takes place within the surface film and within only a few nanometres of the surface of the substrate. Under these conditions, the scattering of the incident radiation by the substrate is much reduced and detection limits of the order of 10^{10} atoms cm^{-2} are possible. The technique has many applications in the study of surface layers, particularly when coupled with measurements of the specularly and nonspecularly reflected components of the incident beam. By varying the angle of incidence, the depth of penetration can be varied which allows information about the composition of multilayers to be obtained. A recent review[20] gives a very clear account of the development of the technique, the underlying principles and its application.

An account of the fundamentals and practice of XRF analysis up to the early 1970s is given by Jenkins[21]; for an extensive review of the fundamentals and the application to silicate rock analysis the reader is referred to the 1987 handbook by Potts[22].

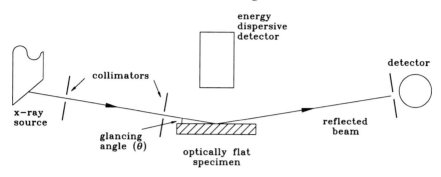

Figure 12. Arrangement for total reflection XRF. The detector at the right of the diagram allows measurements of reflectivity to be made.

At the present time, much effort in the development XRF analysis is directed toward selected area microanalysis using synchrotron radiation. This work is described in a later section.

Localised Analysis and the Electron Microprobe

Almost all of the measurements made during the interwar period were bulk analyses in which no information on distribution was obtained other than by mechanical separation of the analysed sample. One exception is the work of von Hámos in Sweden[23,24] who first used the term 'x-ray microanalyser' to describe an apparatus for producing images showing the distribution of elements at the surface of polished specimens. The geometry of his curved crystal point to point imaging device is illustrated in figure 13 and examples of its application in figure 14. The spatial resolution appears to be of the

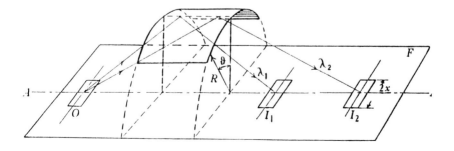

Figure 13. Basic geometry of x-ray microanalyser, or imaging spectrograph[24].

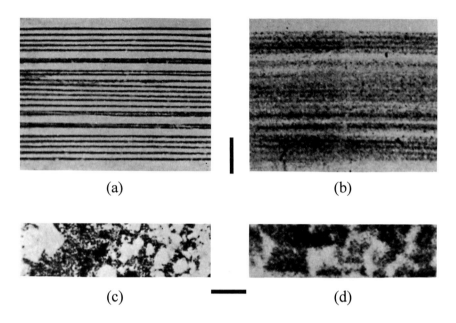

(a)   (b)

(c)   ▬   (d)

Figure 14. Examples of images obtained with the imaging spectrograph: a) test specimen with thin plated layers of iron and b) the x-ray image in Fe K_α radiation; c) a polished specimen of an ore and d) its x-ray image in Fe K_α radiation showing the distribution of iron[49]. Both scale bars represent 1 mm.

order of 50 μm but it is clear that problems could arise due to overlapping images from elements with closely spaced emission lines and that quantification would involve accurate densitometry of the images.

After the Second World War, Engström's school at the Karolinska Institute in Stockholm developed techniques for the quantitative densitometry of contact microradiographs[25]. This technique was applied by Lindström[26,27] to quantitative analysis of inorganic elements in biological tissue by differential absorption measurements at wavelengths on either side of absorption edges.

The first development of a versatile x-ray microanalyser in which elemental distribution of a wide range of elements could be studied with a high spatial resolution did not occur until 1949, although an earlier patent by Hillier[28] described an instrument that was never constructed. Raymond Castaing, a student of Guinier in the University of Paris, converted an electron microscope to operate as an electron probe[29,30] in which electrostatic lenses were used to focus an electron beam to a diameter of about 1 µm at the surface of the specimen. X-rays emitted from the point of impact of the probe were analysed by a Bragg spectrometer using a curved crystal in order to increase the efficiency of collection of x-rays. The intensity was measured by an electrical detector so that a readout could be obtained in a few seconds. An optical microscope mounted coaxially with the electron beam and an *x-y* stage holding the specimen and standards allowed the operator to select at will any area for analysis (figures 15 and 16).

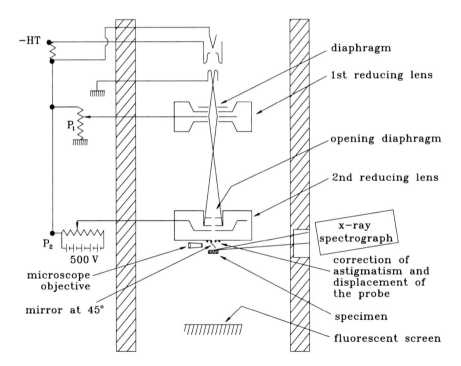

Figure 15. Diagrammatic representation of Castaing's electron microprobe. The beam from the electron gun at the top of the evacuated column is focused by two electrostatic lenses to form a demagnified image of the source at the surface of a specimen mounted on a stage with orthogonal traverses. The specimen is viewed with a microscope via a 45° mirror with a central hole to permit passage of the electron beam. The crystal spectrometer records the emitted x-ray spectrum[30]. Almost all subsequent electron probe instruments have used magnetic lenses.

Figure 16. Castaing's electron microprobe[30].

Castaing's instrument was truly a 'microanalyser': the volume irradiated by the electron beam was of the order of a few cubic micrometres weighing some 10^{-11} g, so that for a minor element present at 100 ppm by weight the actual mass of emitting element was in the region of 10^{-15} g.

Castaing not only developed the instrument but also showed how it was possible to calculate the effect of one element upon the measured characteristic intensity of another in the analysed volume by semi-empirical modelling of the processes involved in the generation of the x-rays and their escape from the specimen. This was of great practical importance since it permitted quantitative analysis with standards whose matrix composition was different from that of the specimen. It made it possible to use a piece of pure metallic iron as a standard for determining the concentration of iron in, for example, a silicate slag. There were limitations to the accuracy of such calculations and today, some 45 years on, the subject of matrix corrections is still an active field, particularly in the case of the light elements whose emissions are of low energy. Castaing's microprobe was used in quantitative analysis by a number of workers in France, notably in the study of intermetallic diffusion and the composition of intermetallic phases formed in diffusion couples[31–33].

Several experimental electron probe instruments were developed in the 1950s following Castaing's original work but it was not until 1960 that commercial instruments began to appear. In France, the Cameca Company developed an instrument based on Castaing's design, while in England the Cambridge Instrument Company marketed the 'Microscan I' based closely on a design by Peter Duncumb and David Melford[34]. From 1960 to 1965 I was involved in the design of the 'Geoscan', particularly for the analysis of minerals in petrological material. A.E.I. Scientific Instruments produced an instrument primarily for metallurgical material which followed closely the prototype constructed by Tom Mulvey[35]. In the USA, Applied Research Laboratories collaborated with David Wittry in the design of the ARL EMX instrument, while in Japan the JEOL Company marketed an instrument very similar to the first Cameca design. Nowadays the principal manufacturers selling fully equipped electron probe microanalysers are Cameca and JEOL although, as discussed below, many analyses are carried out in instruments initially designed as scanning or transmission electron microscopes.

Beam Scanning

It is impossible to dissociate the history of the electron microprobe from that of the scanning electron microscope (SEM), first developed as a high resolution instrument by McMullen and Oatley[36]. In the SEM, the electron beam is scanned over a small area of the specimen in a television like raster. A magnified image is produced by collecting secondary electrons and using the signal to control the brightness of the spot on a cathode ray tube scanned in synchronism with the microscope beam.

While Castaing was completing his electron probe in Paris, the prototype SEM was being built in the Engineering Department at Cambridge University and Bill Nixon was working a few hundred yards away in the Cavendish Laboratory on the point projection x-ray microscope[37]. All three instruments, developed quite independently, used essentially identical electron optical systems. It was therefore a very natural step for Ellis Cosslett, in charge of the Electron Microscope Section in the Cavendish, to set one of his research students the task of building a scanning x-ray microscope. When this proved to have few advantages over the static instrument, the student, Peter Duncumb, suggested that a scanning electron probe, with the image brightness controlled by a selected x-ray signal, would be a better proposition. So it proved, and such was the speed with which a homemade instrument could be modified it was but a matter of days before pictures were obtained (figure 17)[38,39].

The combination of high resolution mapping of element distributions with the ability to stop the scanning beam on any selected

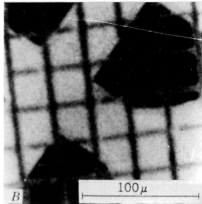

Figure 17. Scanning images of superimposed grids of copper and silver obtained by energy selection with a proportional counter; a) silver L radiation and b) copper K radiation[38].

feature for detailed quantitative analysis opened up almost limitless fields of investigation in materials science and in the earth sciences. I had the privilege to be a member of the Cavendish group at the time and built a simple scanning generator to fit on a microprobe constructed for examining minerals and rocks. Figure 18 shows the first scanning images obtained on this type of material[40,41]. Superimposed on the photographs, by a simple double exposure of the CRT display, are the traces showing semi-quantitatively the distribution of elements between different phases.

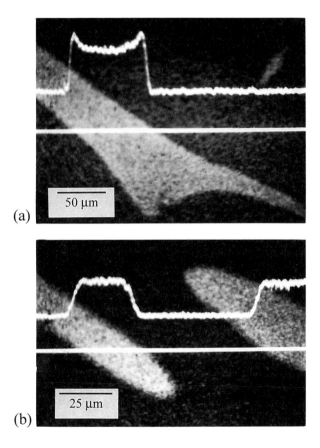

(a)

(b)

Figure 18. X-ray scanning images showing a) the distribution of nickel in kamacite and taenite of the iron meteorite, Canyon Diablo, and b) the distribution of calcium between augite and hypersthene in inverted pigeonite from the Bushveld intrusion in South Africa. Superimposed traces show variation of nickel K_α and calcium K_α along the marked tracks of probe[40,41].

Figures 19a and 19b illustrate another early example of the use of scanning x-ray images in metallurgy in which an important defect in worked steel billets known as 'hot shortness' was shown to be due to the presence of nickel, copper and tin remaining at the surface of the metal under the oxide scale during prolonged heating[42]. In further work on this problem, Melford[43] used both scanning images (figures 19c and 19d) and quantitative point analysis to elucidate the role of grain boundary diffusion in dispersing concentrations of copper and nickel below the surface of the oxide scale. Scanning pictures revealed concentrations of nickel and copper at points in the grain boundaries below the surface of oxidised steel. This behaviour, which appeared to indicate 'uphill' diffusion, was attributed to the presence of low energy sites, possibly associated with triple junctions (figure 19e). Samples of mild steel, plated with copper and nickel, were also used to study relative grain boundary diffusion rates and, for the first time, *in situ* heating experiments were made using a microfurnace on the specimen stage. These experiments represent good examples of investigations which would have been much more difficult without the facility of beam scanning.

Application of the technique in biology and medicine was slower to develop, partly because only a few inorganic elements such as calcium and phosphorus are present in tissues in sufficient concentration to bring them within the operating range of the electron probe. Also, the very variable density of biological material and its sensitivity to the electron beam make quantitative interpretation much more difficult. Nevertheless, Tousimis[44] demonstrated the concentration of copper in eye tissues in Wilson's disease. A significant advance in the problem of variable density of the experimental material was made by Hall and Gupta[45] who used the intensity of the continuum as a measure of the mass thickness of the specimen at the point of analysis.

For some years, the electron microprobe and the scanning electron microscope developed as separate instruments; the former characterised by its x-ray spectrometer and optical microscope for location of the beam with a resolution limited to about 1 μm by electron scattering in the specimen. Not all instruments were equipped with beam scanning. The thrust of development in the

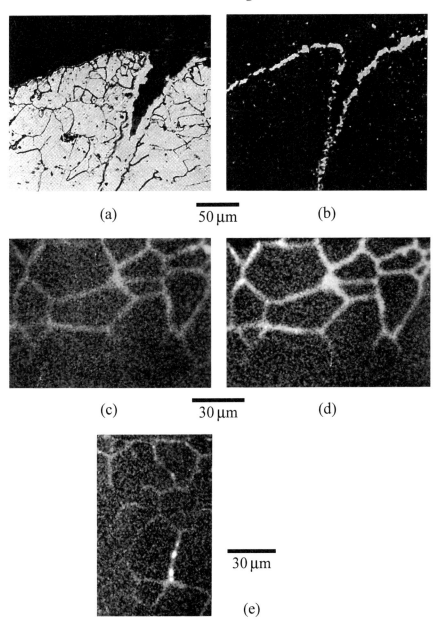

Figure 19. a) Optical micrograph of crack in the surface of a hot worked steel billet; b) scanning image in nickel K_α showing concentration of nickel along the crack[42]; c), d) scanning images showing concentrations of copper and nickel in grain boundaries below the surface; e) scanning image showing nickel concentration at a preferred site below the surface[43]. In each image the surface is at the top.

SEM, on the other hand, was towards high resolution imaging at the 100 Å level with secondary electrons, and the use of other secondary emissions, e.g., light and backscattered electrons as alternative signals for image formation. The distinction between the two instruments has gradually become blurred and although, at the present time, commercial SEMs are still dedicated to operation as versatile imaging instruments, they are designed so that x-ray spectrometers may be fitted as standard accessories. Conversely, electron microprobes are able to operate as SEMs; they are equipped with good optical microscopes as standard features but not with the complex tilting and rotating stages which are essential for microscopy but generally an embarrassment for quantitative analysis.

Energy Dispersive Spectrometers

One of the major factors which brought about the union of the electron probe and the SEM was the development of the so called 'energy dispersive' x-ray detector. The crystal spectrometer, as used in the electron microprobe, is a single channel instrument of some mechanical complexity and the number of elements that can be analysed simultaneously is limited by the physical space available for additional spectrometers and, not least, by their cost.

An energy sensitive detector, such as a proportional counter, provides a potential alternative to the crystal spectrometer. Proportional counters were in use in the 1950s for x-ray detection[46,47], but again as single channel devices, i.e., only one group of output pulses, corresponding to one x-ray wavelength, was recorded at any one time. In the early 1960s Ray Dolby (now well known for his work on noise reduction in audio recording) used proportional counters to analyse the spectra of the light elements more efficiently than was possible with a crystal spectrometer. He devised a multichannel system in which analogue circuitry performed a spectrum deconvolution and generated output voltages proportional to the intensities of individual characteristic x-ray emissions from the specimen. This equipment worked fast enough to allow recording of scanning images[48] of elements down to beryllium ($Z=4$).

Although proportional counters are able to operate at high counting rates and with good signal to noise ratio, they have a rather poor energy resolution so that sensitivity is severely limited when a weak line in the spectrum is overlapped by a neighbouring intense peak. The real impact of energy dispersive spectrometry came with the development of the solid-state silicon x-ray detector whose energy resolution is some 5–10 times better than that of the proportional counter in the x-ray region. This device was first applied in x-ray analysis with the electron microprobe by Fitzgerald, Keil and Heinrich in 1968[49]. The advantage over the proportional counter is shown by figures 20 and 21 which show the spectra recorded by Dolby with a gas filled proportional counter from beryllium, carbon and oxygen and the same peaks (from a sample containing all three elements) recorded with a modern silicon detector coupled to a state of the art amplifier and recording system[50].

The wavelength region commonly used for analysis is from about 12 Å (1 keV) to 1.3 Å (10 keV), including the K lines of the elements from sodium ($Z=11$) to zinc ($Z=29$) and the L and M lines of heavier elements. Here, the energy resolution of the silicon energy dispersive detector, although still not approaching that of the crystal spectrometer, is nevertheless adequate for quantitative analysis of many specimens. In the case of elements present below about 1% by weight, the comparatively low peak to background ratio obtainable

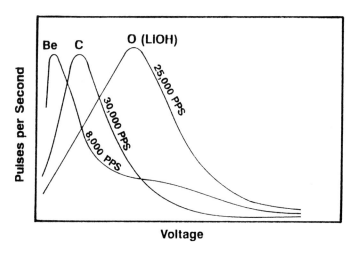

Figure 20. Spectra of beryllium K, carbon K and oxygen K radiations recorded with a gas flow proportional counter[48].

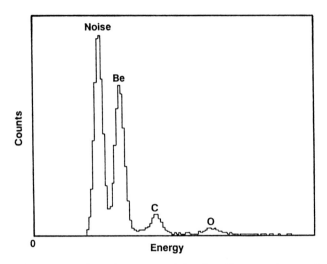

Figure 21. Spectrum of specimen containing beryllium, carbon and oxygen recorded with a lithium drifted silicon detector and a low noise preamplifier[50].

(approximately an order of magnitude lower than for a crystal spectrometer) restricts precision and limits the minimum detectable concentration to about 500–1000 ppm. Further, some spectra with very closely overlapping lines cannot be resolved with any degree of success. However, much effort has been expended in the development of methods of spectrum deconvolution[51] and in the design of amplifier and pulse processor equipment for recording spectra which remain undistorted under a wide range of operating conditions. The contribution of the group led by Kandiah at the Atomic Energy Research Establishment, Harwell, in the 1970s is particularly noteworthy in this context[52–54].

One of the great advantages of the silicon detector is that, in conjunction with a multichannel pulse height analyser, it records all wavelengths simultaneously, in contrast to the crystal spectrometer which, as normally operated, is a single channel instrument. The demands upon the operator are thus very small: with a computer based system and appropriate software, pressing the start key presents a spectrum adequate for qualitative identification on the monitor in a few seconds. A complete quantitative analysis for ten or a dozen elements may typically be performed in 1–2 minutes.

A further important characteristic of the solid-state detector is its high collection efficiency. The small tubular housing of the detector, often only 10–12 mm in diameter, is fitted with an end window so that it can be brought close to the specimen, subtending a solid angle of about 0.1–0.3 sr at the point of impact of the electron probe. This compares favourably with a crystal spectrometer where the solid angle may be typically only 0.01 sr. The high collection efficiency has greatly increased the scope of x-ray microanalysis in the SEM where the electron probe current and hence the resultant x-ray intensity are often 100 times lower than in the electron probe. Many hundreds of SEMs are now equipped with silicon detectors and probably account for a substantial fraction of all x-ray microanalysis, albeit that in many cases the SEM is used qualitatively whereas the electron probe is usually set up for accurate quantitative analysis with a combination of wavelength and energy dispersive spectrometry.

Microanalysis in the Transmission Electron Microscope

In the conventional electron microprobe, the resolution is limited by scattering of the electron beam in the specimen, as shown in figure 22. X-rays may be generated anywhere within the pear shaped volume, the size of which depends on the average atomic number of the specimen. To a first approximation, the extreme depth of penetration, p, in micrometres of an electron beam accelerated by a voltage V kV in a specimen of density of ρ gcm^{-3} is given by $p = 2.5 \times 10^{-2} V^2 \rho^{-1}$, the Thomson-Whiddington Law. Thus, at 20 kV and a specimen density of 3 gcm^{-3}, the penetration will be about 3 μm with an analytical resolution of the same order. Resolution can be improved by reducing accelerating voltage but only by using L and M radiations when the voltage is too low to excite K lines. Such soft x-rays are readily absorbed in the specimen and intensities are low. An alternative approach was put forward by Duncumb who used very thin specimens prepared as for transmission electron microscopy. In contrast to the thick specimen situation, figure 22a, very little scattering occurs in the thin specimen, as shown in figure 22b, and the analytical resolution is now determined by the diameter of the electron probe. Duncumb constructed an instrument,

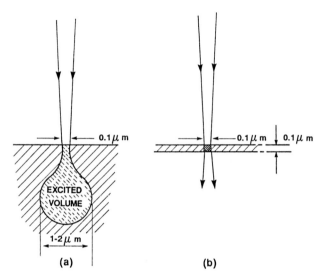

Figure 22. Penetration and scattering of an electron beam in a) a thick specimen and b) a thin specimen.

EMMA, combining transmission electron microscopy and diffraction with microanalysis with crystal spectrometers and used it in studies of grain boundary phenomena[55].

Since that time, microanalytical facilities based on energy dispersive spectrometers have become standard accessories on most transmission microscopes, the high collection efficiency offsetting the low x-ray intensities resulting from low probe currents and the fact that only a small fraction of the incident beam energy is dissipated in the thin section. A recent review of the technique is given by Champness[56].

Localised Trace Element Analysis with X-Ray Excitation

The smallest detectable concentration of any element in the electron microprobe is generally limited, as noted earlier, by the presence of the continuum background. As the concentration is reduced, it eventually becomes impossible to distinguish a characteristic peak from statistical fluctuations in the background. Measurement for long periods gives an improved sensitivity but, ultimately, the long term stability of the incident electron current becomes important and a practical limit of 10–100 ppm, depending on the element and the matrix in which it is contained, is generally realised.

Early attempts to perform microfluorescence analysis were hampered by lack of intensity in the primary source. Zeitz and Baez[57] and Long and Cosslett[58] made use of the very high specific loading of the Cosslett-Nixon x-ray tube by placing apertures close to the thin foil target with the specimen mounted immediately above on a thin plastic film. Although good peak to background ratios could be achieved, signal intensities were low and analysis with micrometre resolution was not feasible. In experiments with single nerve cells[59] it was found to be possible to detect potassium in amounts of the order of 0.1–1 ng in areas of the order of 50 μm diameter.

Some improvements in this performance would now be possible with modern silicon detectors. The re-emergence of the technique has, however, been due to the development of the electron synchrotron as a radiation source able to produce x-ray beams with intensities many orders of magnitude greater than from any x-ray tube. A small probe at the specimen may be produced by aperturing the synchrotron beam or by focusing with a combination of curved Bragg monochromators and grazing incidence, total reflection mirrors. In addition to the very high available intensity, microanalysis benefits from the fact that the radiation is polarised. By collecting fluorescence x-rays from the sample at right angles to the plane of polarisation the background due to scattering is much reduced.

Synchrotron radiation x-ray fluorescence analysis (SR–XRF) has a very wide potential, particularly with the increasing intensities available from new generation sources. Quantitative trace element analysis has long been an important technique in the earth sciences and SR–XRF with a collimated or focused probe provides one way of extending the range of area selected analysis to concentrations beyond the reach of the electron probe. Detection limits vary between different experimental arrangements and, for a fixed recording time, with the spatial resolution required. With a 20 μm×20 μm spot on sulphide minerals in coals, detection limits of 5–50 ppm for elements with $Z>25$ have been obtained[60]. Bearing in mind that these figures relate to recording with a silicon detector, they represent a substantial improvement over the electron probe where, with the same detector, limits would be closer to 500 ppm. Just as with the electron probe, a significant improvement would be obtained with a crystal spectrometer, particularly with new, more intense sources to offset the lower collection efficiency.

In addition to its use in measurements of trace element concentrations, the intense continuum of the synchrotron beam also greatly facilitates the recording of the fine structure at x-ray absorption edges (EXAFS) which is related to the environment of the atoms of the element in the absorber. Thus valuable information may be obtained on the chemical nature of particular elements in a small area of a specimen. A recent review of the geological applications of synchrotron radiation by Smith and Rivers[61] provides a detailed assessment of the technique and its capability, together with an extensive bibliography.

The SR–XRF technique has significant advantages for the examination of biological material since important trace elements, present at levels difficult or impossible to detect with the electron probe, can be measured. Although the material is radiation sensitive and some damage is produced by the x-ray beam, the energy deposition is much less than in the electron probe. Moreover, measurements of wet specimens may be made in air or in a helium atmosphere. Many papers, however, present only intensity data with no attempt to convert to elemental concentrations. An example of the sensitivity of the technique is provided by figure 23 which shows the spectrum obtained from a single liver cell with a beam 80–100 μm in diameter and with a measuring time of 40 minutes[62].

Localised Trace Element Analysis with Proton Excitation

Characteristic x-ray emission is also produced when targets are bombarded with protons. Although the efficiency of x-ray production is much lower than for electrons, this is more than compensated by a very low bremsstrahlung background with the result that detection limits, like those for SR–XRF, are considerably below those of the electron probe.

Analytical use of proton excitation for analysis (proton excited x-ray emission or PIXE) was first described in 1970[63]. A broad

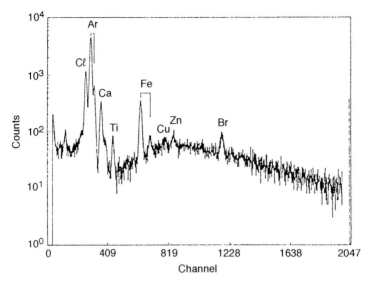

Figure 23. Synchrotron XRF spectrum from a single liver cell[62]. The argon peak suggests that the spectrum was recorded in air.

proton beam was used to analyse the deposit on a carbon foil which had been exposed to the atmosphere for several hours (figure 24). Although collimated proton beams of low intensity had been recorded by a number of authors, practical localised analysis was made possible by the use of magnetic quadrupole lenses to focus the proton beam to a small probe. Rotationally symmetric magnetic lenses of the type used in the electron microprobe are far too weak to focus the 2–6 MeV protons required for useful x-ray production and electrostatic lenses would require almost impossibly high electrode potentials. Strong focusing magnetic quadrupoles, however, are well suited to this purpose. A single quadrupole acts as a cylindrical lens and produces an elongated image from a round object; a combination of two, three or four lenses may be used to give focusing in two planes and to image a round source as a demagnified round image. The first probe system of this type was described in 1970 by Cookson and Pilling[64], working at AERE Harwell in the UK. This apparatus produced a minimum probe diameter of $3\,\mu m$ with a current density of about $150\,pA\,\mu m^{-2}$. The development of nuclear microprobes, their optics and application has been reviewed in detail

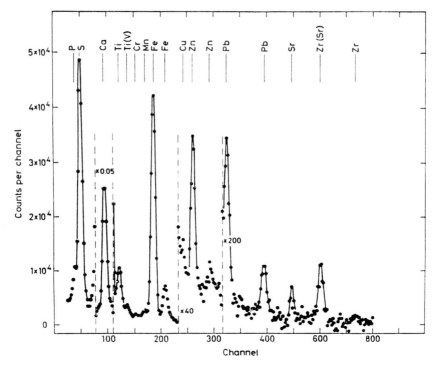

Figure 24. X-ray spectrum from dust particles collected on a carbon film and excited by proton bombardment[63].

by Cookson[65] and the progress of the technique may be followed in the proceedings of international conferences on PIXE and its analytical applications[66]. The technique has found wide application in the analysis of particulates and aerosols and in the analysis of geological material[67]. Figure 25 illustrates the greatly improved peak to background ratio and sensitivity for trace elements compared with the electron probe.

The potential in the biological field would seem to be considerable but the view of Johansson, one of the pioneers of the technique, is interesting in this context[68]: "Although quite a few interesting results have been obtained, it is difficult to discern any real breakthrough in this field". Johansson attributes this state of affairs partly to the difficulty of performing meaningful investigations on biochemical processes, as compared with the simple procedures for analysing geological material or particulates. Further, in the analysis

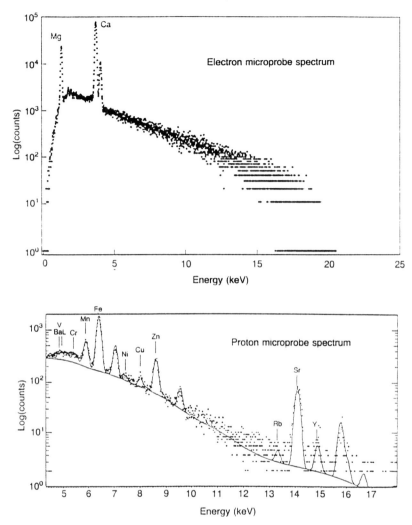

Figure 25. A comparison of the x-ray spectra from a dolomite sample observed using electron and proton probes. The curve for the proton probe was severely attenuated for energies below about 4.5 keV to reduce the x-ray counts from magnesium and calcium, the major elements present[67].

of inorganic materials the general nature of the science is such that it is usually possible for the analyst to acquire sufficient expertise in the appropriate field in order to carry out useful studies. The complexity of biomedical research, however, is such that this overlap of expertise is much less likely and important advances are then dependent upon a successful collaboration between scientists in two widely different fields. This problem is not, of course, confined to PIXE analysis.

Nevertheless, examples of such collaboration do exist, as illustrated by the SR–XRF analysis shown in figure 23. Again, in figure 26, scanning PIXE analysis has been used in a collaborative study of the distribution of platinum and iodine containing anti-cancer drugs in a tumour cell[69]. Platinum appears to be uniformly distributed over the whole cell while iodine and iron are spatially correlated in what is assumed to be the region of the nucleus. These two-dimensional distributions, incidentally, illustrate well the problem of interpreting absolute concentrations from x-ray intensity measurements on biological tissue: the carbon distribution gives a measure of the mass thickness of organic material to which the remaining x-ray distributions must be related to give absolute concentrations.

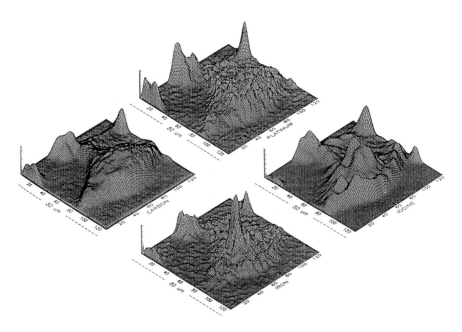

Figure 26. PIXE microanalysis of a tumour cell after multidrug exposure to cisplatin and iododeoxyrubicin. Strong spatial correlation of iron and iodine occurs in the assumed nuclear region. Platinum is homogeneously distributed in the whole cell[69].

Current Activity

The electron probe and related techniques of direct electron excitation of the specimen have reached a degree of maturity in the 45 years since Castaing's initial experiments. Electron optical components have improved in design and brighter electron sources such as lanthanum hexaboride and field emitters have made it possible to form small probes with current densities several orders of magnitude higher than normally employed in the conventional electron microprobe. The physical limitations on resolution imposed by electron optics, scattering in the specimen and, not least, by beam damage to the specimen are reviewed by Champness[56]. It is possible to make analyses on areas of a few tens of nanometres in diameter using thermionic sources and the minimum detectable mass of an element can be as low as 10^{-20} g, corresponding to a few hundred atoms. With field emitting guns, compositional changes over a few nanometres can be recorded and detection of single atoms is feasible.

In conventional electron probe analysis, attention is being directed to improving accuracy of quantitative measurements, particularly for the light elements, and to the analysis of specimens with rough surfaces where absorption and other corrections cannot be calculated accurately. The analysis of layered specimens with the penetration of the beam varied by the use of different accelerating voltages has been explored and modelled by Monte Carlo calculations of single electron trajectories[70]. The Monte Carlo method, first introduced in the modelling of x-ray production over 30 years ago by Green[71], is still widely used and, with increased computing power, is now an even more powerful tool in this field.

One area in which significant development may be expected is x-ray spectrometry, which is basic to all forms of x-ray microanalysis. At best, we only use some 5% of the available x-ray emission and often a much smaller fraction. The resolution of silicon detectors is limited by the intrinsic properties of the silicon and also by the noise introduced by the preamplifier. An improvement in the latter has resulted from the development of a five electrode FET preamplifier[50] and the use of germanium instead of silicon which, with a smaller bandgap, gives a smaller natural pulse height distribution from the detector itself. Possibilities also exist for improvements in crystal spectrometers to give parallel detection of a number of wavelengths, although Wittry[72] concludes that, unless it is essential to measure a number of elements simultaneously, the parallel detection spectrometer will not compare in performance with a scanning monochromator under computer control. The development of multilayer monochromators and total reflection mirror systems is also directly relevant to the focusing of synchrotron beams for the formation of small probes.

Whatever new developments lie over the horizon, and despite the many new analytical techniques which have appeared since the Second World War, x-ray microanalysis in its various forms seems likely to remain perhaps the most powerful of all methods of elemental analysis because of the unique and simple relationship of characteristic x-ray emission to atomic structure that Moseley had elucidated within eighteen years of Röntgen's original discovery. It is impossible to predict how the second 100 years will compare in excitement with the first, but the ever widening range of applications and the instrumental developments of the last two decades alone will ensure that the next 50 years, at least, will not be dull.

Acknowledgements

I am most grateful to Dr S J B Reed for discussions in the course of preparing this paper and to Mr A Iredale for help in assembling the figures.

References

(1) H G J Moseley "The High Frequency Spectra of the Elements" *Phil. Mag.* **26** 1024 (1913)

(2) H G J Moseley "The High Frequency Spectra of the Elements; Part II" *Phil. Mag.* **27** 703 (1914)

(3) W Friedrich *et al* "Interferenzersheinungen bei Röntgenstrahlen" *Ann. der. Phys.* **41** 971 (1913)

(4) M von Laue "Interferenzerscheinungen bei Röntgenstrahlen" *Ann. der Phys.* **41** 989 (1913)

(5) C G Barkla & C A Sadler "Homogeneous Secondary Röntgen Radiations" *Phil. Mag.* 6th series **16** 550 (1908)

(6) W H Bragg & W L Bragg "The Reflection of X-Rays by Crystals" *Proc. Roy Soc.* **A88** 429 (1913)

(7) G W C Kaye "The Emission and Transmission of Röntgen Rays" *Phil. Trans. Roy. Soc.* **A209** 123 (1909)

(8) G de Hevesy "Chemical Analysis by X-Rays and its Applications" New York: McGraw Hill (1932)

(9) A Alexander & A Faessler "Eine neuelichtstarke Röntgenrohre für Fluoreszenzerregung" *Zeit. für Physik* **68** 260 (1931)

(10) M Siegbahn "The Spectroscopy of X-Rays" Oxford: Oxford University Press (1925)

(11) D Coster & G de Hevesy "On the Missing Element of Atomic Number 72" *Nature* **111** 79 (1923)

(12) H Friedman & L S Birks "A Geiger Counter Spectrometer for X-Ray Fluorescence Analysis" *Rev. Sci. Instrum.* **19** 323 (1948)

(13) E T Hall "Some Uses of Physics in Archaeology" *Year Book of the Physical Society (London)* p 22 (1958)

(14) F Claisse "Accurate X-Ray Fluorescence Analysis without Internal Standard" *Preliminary Report No. 327 Department of Mines Quebec.* (1956)

(15) K Norrish & J T Hutton "An accurate X-Ray Spectrographic Method for the Analysis of a Wide Range of Geological Samples" *Geochim. Cosmochim. Acta* **33** 431 (1969)

(16) I Adler & J M A Axelrod "A Multiwavelength Fluorescence Spectrometer" *J. Opt. Soc. Amer.* **43** 769 (1953)

(17) V Mazo-Gray & M Alvarez "X-Ray Fluorescence Characterization of Ming Dynasty Porcelain Rescued from a Spanish Shipwreck" *Archaeometry* **34** 37 (1992)

(18) Y Yoneda & T Horiuchi "Optical Flats for Use in X-Ray Spectrochemical Analysis" *Rev. Sci. Instrum.* **42** 1069 (1971).

(19) H Aiginger & P Wobrauschek "A Method for Quantitative X-Ray Fluorescence Analysis in the Nanogram Region" *Nucl. Instrum. Meth.* **114** 157 (1974)

(20) D K G de Broer *et al* "Glancing-Incidence X-Ray Analysis of Thin Layered Materials. A Review" *X-Ray Spectrometry* **24** 91 (1995).

(21) R Jenkins "An Introduction to X-Ray Spectrometry" London: Heyden (1974)

(22) P J Potts "A Handbook of Silicate Rock Analysis" Glasgow: Blackie (1987)

(23) L von Hámos "Microchemical Analysis of Plane Polished Surfaces by means of Monochromatic X-Ray Images" *Nature* **134** 181 (1934)

(24) L von Hámos "X-Ray Microanalyser" *J. Sci. Instrum.* **15** 87 (1938)

(25) O Hallén & H Hydén "Quantification and Automation in Contact Microradiography" *X-Ray Microscopy and Microradiography* (V E Cosslett *et al*, eds) New York: Academic Press p 249 (1957)

(26) B Lindström "Quantitative Cytochemical Elementary Analyses by X-Ray Absorption Spectrophotometry" *X-Ray Microscopy and Microradiography* (V E Cosslett *et al*, eds) New York: Academic Press p 443 (1957)

(27) B Lindström "Röntgen Absorption Spectrophotometry in Quantitative Cyto-chemistry" *Acta Radiologica Suppl.* **125** (1955)

(28) J Hillier *US Patent 2418029* (1947)

(29) R Castaing & A Guinier "The Application of Electron Probes to Metallurgical Analyses" *Proc 1st. Int. Conf. Electron Microscopy* p 60 (1950)

(30) R Castaing "Application des Sondes Electroniques a une Methode d'Analyse Ponctuelle Chimique et Crystallographique" *PhD Thesis, Univ. of Paris* (Onera Publ. No. 25) (1951)

(31) J Philibert & Y Adda "On Establishing Equilibrium Diagrams of Binary Alloys by Observations of Intermetallic Diffusion. Application to the Uranium-Zirconium System" *C. R. Acad. Sci. Paris* **245** 2507 (1957)

(32) J Philibert & H Bizouard "Some New Applications of Castaing's Electron Microprobe and their Practical Importance" *Rev. Metall.* **56** 187 (1959)

(33) J Philibert "The Castaing Microsonde in Metallurgical and Mineralogical Research" *J. Inst. Metals* **29** 241 (1962)

(34) P Duncumb & D A Melford "Design Considerations of an X-Ray Scanning Microanalyser used mainly for Metallurgical Applications" *X-Ray Microscopy and Microanalysis* (A Engström *et al*, eds) Amsterdam: Elsevier p 358 (1960)

(35) T Mulvey (1960) "A New Microanalyser" *X-Ray Microscopy and Microanalysis* (A Engström *et al*, eds) Amsterdam: Elsevier p 372 (1960)

(36) D McMullen "An Improved Scanning Electron Microscope for Opaque Specimens" *Proc. IEE* **100** 254 (1953)

(37) V E Cosslett & W C Nixon "X-Ray Shadow Microscope" *Nature* **168** 24 (1951)

(38) P Duncumb & V E Cosslett "A Scanning Microscope for X-Ray Emission Pictures" *X-Ray Microscopy and Microradiography* (V E Cosslett *et al*, eds) New York: Academic Press p 374 (1957)

(39) P Duncumb "The X-Ray Scanning Microanalyser" B*rit. J. Appl. Phys.* **10** 420 (1959)

(40) J V P Long "Microanalysis with X-Rays" *PhD Thesis, Univ. of Cambridge* (1958)

(41) S O Agrell & J V P Long "The Application of the Scanning X-ray Microanalyser to Mineralogy" *X-Ray Microscopy and Microanalysis* (A Engström *et al*, eds) Amsterdam: Elsevier p 391 (1960)

(42) D A Melford "The Application of an Improved X-Ray Scanning Microanalyser to Problems in Ferrous Metallurgy" *X-Ray Microscopy and Microanalysis* (A Engström *et al*, eds) Amsterdam: Elsevier p 407 (1960)

(43) D A Melford "A Study of Microsegregation at Grain Boundaries in Mild Steel by Means of the Electron Probe Microanalyser" *X-Ray Optics and X-Ray Microanalysis* (H H Pattee *et al*, eds) New York: Academic Press p 577 (1963)

(44) A J Tousimis "Electron Probe Microanalysis of Biological Specimens" *X-Ray Optics and X-Ray Microanalysis* (H H Pattee *et al*, eds) New York: Academic Press p 539 (1963)

(45) T A Hall & B L Gupta "The Localization and Assay of Chemical Elements by Microprobe Methods" *Quart. Rev. Biophys.* **16** 279 (1983)

(46) A R Lang "The Proportional Counter in X-ray Diffraction Work" *Nature* **168** 907 (1951)

(47) U W Arndt *et al* "A Gas Flow Proportional X-Ray Diffraction Counter" *Proc. Phys. Soc.* **B67** 357 (1954)

(48) R M Dolby "An X-Ray Microanalyser for Elements of Low Atomic Number" *X-Ray Optics and X-Ray Microanalysis* (H H Pattee *et al*, eds) New York: Academic Press p 483 (1963)

(49) R Fitzgerald *et al* "Solid-State Energy Dispersive Spectrometer for Electron Probe X-Ray Analysis" *Science* **159** 528 (1968)

(50) T Nashabishi & G White "A Low-Noise FET with Integrated Charge Restoration for Radiation Detectors" *IEEE Trans. Nucl. Sci.* **37** 452 (1990)

(51) S J B Reed "Electron Microprobe Analysis" (2nd Ed) Cambridge: Cambridge University Press (1993)

(52) K Kandiah "Semiconductor Nuclear Particle Detectors and Circuits" *National Academy of Sciences Publication 1593* (L Brown *et al*, eds) p 495 (1969)

(53) K Kandiah "High Resolution Spectrometry with Nuclear Detectors" *Nucl. Instr. Methods* **95** 289 (1971)

(54) K Kandiah *et al* "A Pulse Processor for X-Ray Spectrometry with Si-Li Detectors" *Proc. 2nd. ISPRA Nucl. Elec. Symposium (EUR 5370C)* Luxembourg: Commission of Eur. Comm. Directorate p 513 (1975)

(55) P Duncumb "X-Ray Microanalysis of Elements in the Range $Z=4-92$ Combined with Electron Microscopy and Electron Diffraction" *X-Ray Optics and X-Ray Microanalysis* (H H Pattee *et al*, eds) New York: Academic Press p 431 (1963)

(56) P Champness "Analytical Electron Microscopy" *Microprobe Techniques in the Earth Sciences. The Mineralogical Society Series 6* (P J Potts *et al*, eds) London: Chapman and Hall p 91 (1995)

(57) L Zeitz & A V Baez (1957) "Microchemical Analysis by Emission Spectrographic and Absorption Methods" *X-Ray Microscopy and Microradiography* (V E Cosslett *et al*, eds) New York: Academic Press p 417 (1957)

(58) J V P Long & V E Cosslett "Some Methods of X-Ray Microchemical Analysis" *X-Ray Microscopy and Microradiography* (V E Cosslett *et al*, eds) New York: Academic Press p 435 (1957)

(59) J V P Long & H O E Röckert "X-Ray Fluorescence Microanalysis and the Determination of Potassium in Nerve Cells" *X-Ray Optics and X-Ray Microanalysis* (H H Pattee *et al*, eds) New York: Academic Press p 513 (1963)

(60) R M White *et al* "Analysis of Iron Sulphides from UK Coal by Synchrotron X-Ray Fluorescence" *Fuel* **68** 1480 (1989)

(61) J V Smith & M L Rivers "Synchrotron X-Ray Microanalysis" *Microprobe Techniques in the Earth Sciences. The Mineralogical Society Series 6* (P J Potts *et al*, eds) London: Chapman and Hall p 161 (1995)

(62) H Yao *et al* "Element Distributions and Quantitative Analysis of a Single Cell by Micro-PIXE and Synchrotron XRF" *Nucl. Instrum Methods* **B75** 563 (1993)

(63) T B Johansson *et al* "Elemental Trace Analysis at the 10^{-12} g Level" *Nucl. Instr. Methods* **84** 141 (1970)

(64) J A Cookson & F D Pilling "A 3 MeV Proton Beam of less than 4 Microns Diameter" *Harwell Report AERE-R* 6300 (1970)

(65) J A Cookson "The Production and Use of a Nuclear Microprobe of Ions at MeV Energies" *Nucl. Instrum. Methods* **165** 477 (1979)

(66) M Uda (ed) "Proc. Sixth Int. Conf. on PIXE" *Nucl. Instrum. Methods* **B75** (1993)

(67) D G Fraser "The Nuclear Microprobe — PIXE, PIGE, RBS, NRA and ERDA" *Microprobe Techniques in the Earth Sciences. The Mineralogical Society Series 6* (P J Potts *et al*, eds) London: Chapman and Hall p 141 (1995)

(68) S A E Johansson "Summary of 6th Int Conf on PIXE and its Applications" *Nucl. Inst. Methods* **B75** 589 (1993)

(69) P Moretto *et al* "PIXE Microanalysis in Human Cells: Physiology and Pharmacology" *Nucl. Inst. Methods* **B75** 511 (1993)

(70) N Ammann & P Karduck "A Further Developed Monte Carlo Model for the Quantitative EPMA of Complex Samples" *Microbeam Analysis — 1990* (J R Michael and P Ingrams, eds) San Francisco: San Francisco Press p 150 (1990)

(71) M Green "Monte Carlo Calculations of Spatial Distribution of Characteristic X-Ray Production in a Solid Target" *Proc. Phys. Soc.* **83** 204 (1963)

(72) D Wittry "New Doubly Curved Diffractor Geometries and their Use in Microanalysis" *X-Ray Optics and Analysis: Inst. Phys. Conf. Series 130* (P B Kenway *et al*, eds) p 535 (1993)

X-Ray Diffraction

Watson Fuller
Keele University

T he immediate popular interest in the discovery of x-rays owed much to the demonstration of their potential for revealing internal structure. The photographs showing the bones in the hand of Röntgen's wife and the arrangement of coins within a wooden box rapidly captured the public imagination. This ability to produce 'shadowgraphs', illustrating the internal structure of a wide variety of animate and inanimate objects, was to find early applications in medicine and a whole range of technologies. However, the spatial resolution of the information obtained in such applications was relatively coarse. Our understanding of the properties of materials, in terms of the three-dimensional structure of the molecules from which they are composed, is dominated by analyses of the scattering of x-rays outside rather than within the geometrical shadow.

Beyond the Geometrical Shadow

The intensity of radiation scattered outside the geometrical shadow is very weak compared to that of the incident x-ray beam and the diffraction effects upon which the visualisation of molecular structure is based were not demonstrated until 1912, almost two decades after Röntgen's discovery. The experiments performed by Friedrich and Knipping, at the suggestion of Max von Laue[1], were aimed primarily at resolving the controversy over whether x-rays were waves or particles. However, in the hands of W L Bragg and his father W H Bragg[2] x-ray diffraction was developed into a technique which has been used to determine structures as simple as common salt and as complex as a virus. It has also been used to characterise types of disorder in materials and to follow transitions in molecular

X-RAYS: The First Hundred Years
Edited by Alan Michette and Sławka Pfauntsch © 1996 John Wiley & Sons Ltd

structure and organisation during biologically and technologically important processes. These achievements have only been possible because of developments in the power of x-ray sources, in beamline optics allowing precise control of incident x-ray wavelength and beam geometry, in electronic area detectors which extend the power of film recording by allowing diffraction data to be displayed in real time and in high speed digital computers which have stimulated new analytical techniques. The success of x-ray diffraction has made three-dimensional molecular structure the natural framework within which fundamental and applied problems in physics, chemistry and biology are analysed.

Any survey of the contribution of x-ray diffraction over the last eighty years to our understanding of the natural world and our capability for changing it must necessarily be highly selective; this review focuses on four themes which are of central importance throughout science and technology. These are: (i) the consequences of regularity and order, (ii) the characteristics of hierarchies, (iii) the interaction between objects and their environment and (iv) the relationship between static structure and dynamic function. One of the most striking aspects of these themes has been the rapidly increasing complexity and sophistication of the models which illustrate them and in which some of the most important contributions of x-ray diffraction were initially formulated.

X-ray diffraction is often considered to be a difficult technique, demanding a high degree of physical and mathematical expertise. Nevertheless it would be remiss not to make an effort to explain the principles of x-ray diffraction. The first part of the review attempts to do this in a way which is free from mathematical formulæ. For those readers who would prefer to concentrate on the results which have been obtained by applying these techniques, the sections from "Highlighting Regularity" onwards are designed to be as self-contained as possible.

The Phase Problem
Figure 1 illustrates the relationship between an object O, its Fraunhofer diffraction pattern DP, and an image of the object I. It shows, in particular, that all points of the object contribute to each point in

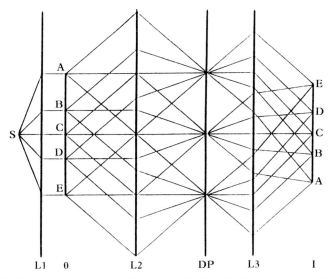

Figure 1. Ray diagram illustrating the relationship between an object O, its Fraunhofer diffraction pattern (at DP) and an image I of the object. L1, L2 and L3 are lenses and S is a point source of illumination. A, B, C, D and E identify corresponding points on object and image.

the diffraction pattern and that all points of the diffraction pattern contribute to each point in the image. The availability of optical lenses able to fulfil the roles of L1, L2 and L3 means that forming an image of an object using visible light is straightforward. However, since the degree of detail observed in an image depends on the wavelength of the radiation, the resolution in an image formed with visible light is limited to about 0.2 μm. Covalent bonds are typically 1000 times shorter than this and, therefore, to obtain information on the arrangement of atoms within molecules it is necessary to use radiation, such as x-rays, with wavelengths of about 1 Å.

The difficulty in exploiting x-rays in this way is that at such wavelengths lenses which fulfil the role of L3 in figure 1 are not readily available. It might be thought that the lack of x-ray lenses to fulfil the roles of L1 and L2 would also be a serious impediment, but in practice this is not the case and there is no difficulty in recording the variation in intensity across the Fraunhofer diffraction pattern. If it were also possible to record the variation in phase across the Fraunhofer diffraction pattern it would be straightforward to replace lens L3 with a computer and calculate an image of the object by Fourier synthesis. This would be analogous to the way by which the Fraunhofer diffraction pattern of an object of known structure can be calculated by Fourier transformation. The Fourier transform of a structure describes the variation in amplitude and phase across the diffraction pattern; the variation in intensity at any point is readily calculated as the square of the amplitude at that point. While direct measurement of the variation in phase is not possible, several ingenious experimental and analytical techniques have been developed to overcome this 'phase problem' in x-ray diffraction.

Although only rarely completely successful on its own, the Patterson synthesis has been one of the most important techniques employed in solving the phase problem. This synthesis does not include any phase information and is readily calculated from the observed variation in intensity across the Fraunhofer diffraction pattern. In the representation of a molecular structure provided by a Patterson synthesis peaks correspond not to the position of atoms but to the termini of interatomic vectors which represent atomic separations in magnitude and direction. Since x-rays are scattered by electrons, the height of an atomic peak in a Fourier synthesis corresponds to the number of electrons in the atom. On the other hand, the height of a peak in a Patterson synthesis corresponds to the number of electrons in the atom at one end of the interatomic vector multiplied by the number in the atom at the other end. The number of peaks in a Patterson synthesis increases rapidly with the number of atoms in the molecule and, in general, can only be readily interpreted when there is a relatively small number of very strongly scattering atoms whose interatomic vectors dominate the synthesis. In favourable circumstances it is possible to translate such interatomic vectors into absolute atomic positions within the structure.

Before describing techniques for solving the phase problem, it will be useful to introduce reciprocity, sampling and symmetry, which are general characteristics in the relationship between an object and its Fourier transform and hence its Fraunhofer diffraction pattern.

Reciprocity is simply illustrated by considering the x-ray diffraction pattern from a crystal of myoglobin shown in figure 2. The

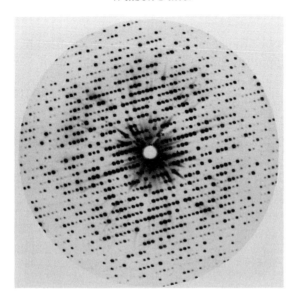

Figure 2. X-ray diffraction pattern from a crystal of the protein myoglobin (courtesy of Sir John Kendrew MRC, Cambridge).

positions of the diffraction spots in this pattern are determined by the arrangement of the molecules of myoglobin in the crystal. As the molecules move further apart the diffraction spots move closer together and *vice versa*. This reciprocal relationship between features in an object and corresponding features in its diffraction pattern is quite general and is reflected in the assumption that the Fourier transform exists in reciprocal space, whereas the object and image exist in real space.

The concept of sampling may be understood by considering the differences between the diffraction pattern from an irregular array of circular apertures with that from a regular array. In the former case the pattern consists of a series of concentric rings which is essentially the same, in terms of the diameter and relative intensity of the various diffraction rings, as that from a single aperture of the same diameter. However, in the latter case diffraction is confined to a series of regularly arranged peaks. The positions and relative intensity of these diffraction peaks can be accounted for by assuming that the intensity distribution in the diffraction pattern from a single aperture has been sampled at positions related to the lattice defining the arrangement of the apertures. This lattice is called the real lattice and the corresponding lattice defining the position of the diffraction peaks is called the reciprocal lattice. The relationship between the real and reciprocal lattices follows the principle of reciprocity de-scribed above — if in the real lattice the distance between neighbour-ing apertures is greater in the vertical than in the horizontal direction then the distance between the peaks in the diffraction pattern will be greater in the horizontal than in the vertical direction. This relation-ship between the two-dimensional real and reciprocal lattices has its parallel in the relationship between the three-dimensional real lattice describing the arrangement of molecules in a crystal and the three-dimensional reciprocal lattice which defines the directions of the x-ray beams diffracted by the crystal. Figure 2 shows an example of such a crystal diffraction pattern — the relative intensities of the various diffraction peaks are determined by the structure of the individual myoglobin molecules while the positions of the peaks are determined by the packing of the molecules. The amplitude and phase of each of the diffracted beams, or reflections as they are called, are combined together in what is known as the structure factor of the reflection.

The symmetry in a circular aperture persists in the circular symmetry of its diffraction pattern. In general, relationships between the symmetry of an object and the symmetry of its Fourier transform, and hence its diffraction pattern, are much more subtle. This is illustrated in figure 3 by the relationship between a model of the DNA double helix and the diffraction pattern given by a fibre made up of DNA molecules with this structure. Because the large number of DNA molecules within the fibre are in a regular array, which can be described by a real lattice, the Fourier transform of a single molecule is sampled at positions determined by the corresponding reciprocal lattice. From the intensity of the diffraction peaks in the centre of the pattern we may deduce that the Fourier transform of a single DNA molecule has a cross like intensity distribution at its centre. Such a distribution is characteristic of helical structures with the angle between the two arms of the cross being related to the angle of climb of the helix. However, if the helix is stretched in the helix

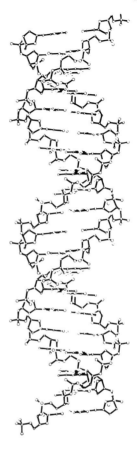

Figure 3. Diffraction pattern from a fibre of DNA (left) and the corresponding model of the DNA (right).

axis direction so that the angle of climb increases, the arms of the cross become more horizontal rather than more vertical. This observation highlights the fact that in the relationship between an object and its diffraction pattern there is not only reciprocity of distance but also reciprocity of direction.

Solutions to the Phase Problem

From the positions of the spots in an x-ray diffraction pattern it is relatively straightforward to determine first the reciprocal lattice parameters and then the corresponding real lattice parameters. For simple structures, such as some crystalline metals, where the object at each lattice point is no more complex than a single atom, it may be possible to determine the complete structure without having to determine the phases of the various structure factors. Also, when the object located at each lattice point, while complex, has a high degree of symmetry, it may be possible to determine reasonably accurate values for the principal parameters defining the structure of the object directly from features in the diffraction pattern. The determination of the double helical structure of DNA by James Watson and Francis Crick[3] using x-ray data recorded by Maurice Wilkins and his collaborators[4,5] is an excellent example of this. The helical character of the structure was revealed by a cross like intensity distribution feature in an x-ray diffraction pattern, similar to but not so well defined as that in figure 3. It was also possible to determine accurately the pitch of the helix and the number of repeating units per pitch and to obtain good estimates for the diameter of the helix and the number of strands per helical molecule. The construction of such a model typically uses information on the chemical structure of the molecule and makes assumptions about covalent bond geometry on the basis of values obtained in simpler compounds. If the model is sufficiently close to the actual three-dimensional structure of the molecule it can be refined by systematic adjustment, based on a least squares or similar procedure, to minimise the discrepancy between the calculated and observed diffracted intensity distributions.

For structures lacking obvious symmetry the derivation of a starting model typically relies more on automatic procedures and less on physical insight and experience in relating features in a diffraction pattern to those in the object. Two approaches of particular importance use the Patterson synthesis as a starting point for locating a strongly scattering atom or atomic grouping which has been introduced as a label into the structure.

In the heavy atom method the starting model is simply the strongly scattering group itself, in the position determined from the Patterson synthesis. The method rests on the assumption that a Fourier synthesis, using the observed structure factor amplitudes and phases calculated simply on the basis of the heavy atom, will show not only this strong scatterer but also indications of atoms in the rest of the structure. These indications provide the basis for improving the model and allowing an 'improved' set of phases to be calculated and hence an improved Fourier synthesis. This process is repeated until no further 'improvement' in the image of the structure takes place and it can be assumed that the complete structure has been revealed. The success of this method depends on the relative scattering power of the heavy atom and the rest of the structure. It works well for small and medium sized structures, e.g., up to molecular weights of about 1000.

In the isomorphous replacement method the change in the diffracted intensities, following binding of a heavy scatterer, is used with knowledge of the position of the scatterer (from a Patterson synthesis) to calculate the phases of the various reflections. This analysis assumes that binding of the heavy scatterer changes neither the original molecular structure nor the crystal packing —hence the name 'isomorphous replacement'. The use of a single isomorphous derivative leaves an ambiguity in the phase determination. Its resolution needs a second isomorphous derivative at a different site before a Fourier synthesis revealing the complete structure can be calculated. The isomorphous replacement method, whose early application before the Second World War was in the determination of the structures of simple inorganic crystals, has been central to the dramatic developments during the last three decades in the determination of the three-dimensional structures of proteins.

During the last decade the determination of the crystal structures of small and medium sized molecules has increasingly become dominated by the determination of phases using so called direct methods. This approach exploits the fact that since the amplitudes and phases of a set of structure factors, describing the diffraction from a particular structure, are both functions of the same electron density distribution there must be mathematical relationships between them. This recognition has resulted in the development of relationships which allow most probable values for the phases of the various structure factors to be derived from the distribution of the amplitudes. Following their success in studies of small and medium sized molecules, these methods are becoming increasingly credible as a way of solving large and complex structures such as proteins. Free of the need to make heavy atom derivatives, direct methods provide a genuine replacement of the lens L3 in figure 1 and represent a true coming of age of x-ray crystallography as a routine technique for visualising molecular structure. Even so, the Fourier synthesis image obtained may need some refinement to optimise the agreement between the calculated and observed structure factor amplitudes.

Highlighting Regularity

A typical sample studied by x-ray diffraction will contain millions of millions of molecules. If these are arranged in a regular array the diffraction from them will, following the previous discussion of sampling, consist of a number of well defined beams. Such beams (or reflections) are a much more striking feature in a diffraction pattern than the diffuse scattering which is characteristic of less well ordered material. It is for these reasons that x-ray diffraction can be said to be a technique which highlights regularity and order at the expense of less regular features. Such highlighting is of particular importance for specimens which, like many industrial polymers and biological materials, consist of crystalline and amorphous regions. Even though the amorphous regions may contain as much scattering material as the crystalline regions, the continuous background which they contribute will appear relatively insignificant because it is distributed widely across the diffraction pattern. This tendency of diffraction techniques to overemphasise the degree of order in materials can be seen as reinforcing general tendencies in perception which highlight symmetry and order. While realising the need to be aware of such tendencies it is also important to recognise that the common occurrence of symmetry and order, in both the animate and inanimate worlds, reflects the functional advantages which such features confer. The development and persistence of such features in living systems is of particular significance, since if they did not offer functional advantages they could not have been expected to survive the competitive pressures of natural selection.

Functional Advantages of Symmetry in Biological Systems

The early x-ray diffraction patterns recorded from fibres of DNA[6] indicated a highly regular structure with a very small repeating unit, i.e., with a volume accommodating less than 100 atoms. Furthermore, DNA from sources as varied as bacteria, plants, fish, rodents and humans gave essentially identical x-ray diffraction patterns. Nor were there significant differences between the diffraction patterns given by DNA from different organs or from normal and cancerous cells. These observations were difficult to reconcile with DNA containing the vast amount of information required to define uniquely each individual organism in terms of the proteins from which it is built up.

This paradox was resolved in 1953 through the proposal by Watson and Crick[3] of the double helical model for the structure of DNA. It can be seen from figure 3 that this model has the overall appearance of a spiral staircase. Within this staircase there are four different types of step. The genetic information stored in a particular molecule of DNA is coded in the sequence of steps, i.e., the message is written in a linear (or one-dimensional) format in a four letter language. During decoding of this information the differences in the detailed chemical structures of the four steps are recognised. However, despite these differences, the geometry of the link between a step and the bannisters of the spiral staircase is identical for all four steps. Thus the external appearence of the DNA double helix can be highly regular and independent of the particular genetic information coded in its chemical structure. It should be added that the structure of the genetic code itself is also highly regular with all the words consisting of three letters, i.e., there are $4^3=64$ distinct words. There are differences in detail between this format for information storage and that characteristic of current information technologies, based on a two letter (binary) code and longer words—typically 8, 16 or 32. However, the use of standardised formats, to allow essentially complete flexibility in information content, is a common feature and is an example of the not infrequent occurrence of nature anticipating technology.

Electron micrographs of the internal structures of cells and other biological entities reveal a high degree of organisation based on regularity and symmetry. Much of this order is beyond the upper size limit for study by current x-ray techniques. However, for a number of viruses detailed structure determination has been possible and has added to our recognition of the ways in which regularity and symmetry are exploited in biological function. The so called spherical viruses, with their blackberry like appearance in the electron microscope, are of particular interest, not least because they include a number of dangerous viruses such as polio and HIV. The icosahedral model for spherical viruses derives from the recognition by Crick and Watson[7] in the 1950s that the regular polyhedra (the Platonic solids) provide a basis for a specific number of identical subunits to occupy equivalent positions. Two polyhedra, the icosahedron and the symmetrically equivalent dodecahedron, have the largest number, 60, of such subunits.

The use of a subunit design has obvious advantages for economy in the genetic information required to code the complete protein coat of the virus. It also increases efficiency in the biochemical process because of the smaller amount of coat material which will have to be rejected as a consequence of an error in protein synthesis. Further, a structure based on identical subunits occupying equivalent positions can also be expected to self assemble in a way which is analogous to crystallisation. The synthesis of the organic molecule urea from manifestly inorganic precursors by the nineteenth century chemist Wohler is often identified as the crucial step in laying to rest vitalist claims that there are fundamental differences between animate and inanimate matter. It would, however, be more accurate to see the assault on vitalism as a continuing one of pushing back boundaries in which each new insight, e.g., an understanding of self assembly, extends the range of biochemical phenomena which can be understood in terms of the laws of physics and chemistry. Recent results in the study of the three-dimensional structures of spherical viruses are described in the section "Relationships Between Structure and Function".

Hæmoglobin, like many proteins with important metabolic roles, consists of not one but a number of protein molecules. Similarities between the structures of these molecules and symmetries in their arrangements confer properties on the complex as a whole beyond those possessed by the individual protein molecules. These properties are reminiscent of the co-operative behaviour exhibited by crystalline arrays. The limited co-operativity exhibited by a symmetrical array of such protein subunits, which individually can assume one of two states, is seen in the 'all or nothing' behaviour which increases discrimination in response to changes in the external environment. Each of the four subunits in hæmoglobin individually undergoes a conformational change following binding of a molecule of oxygen to its hæm group. This conformational change alters the stereochemistry of the contacts between the four subunits. X-ray diffraction studies by Max Perutz and his colleagues[8] showed that these contacts are most favourable if either all or none of the subunits are oxygenated. These observations provided a stereo-chemical basis for the origin of co-operativity in the binding of oxygen—it is stereochemically unfavourable for one subunit to bind oxygen until there is enough oxygen available for all subunits. Similarly, once oxygenated, all of the subunits tend to retain the oxygen bound to them until the level of oxygen in the environment is so low that all of the subunits release their oxygen. This co-operativity confers an enhanced capability for hæmoglobin to bind and retain oxygen in the lungs, where it is abundant, combined with a capacity to release it in the tissues, where it is in short supply.

The Special Case of Mirror Symmetry
It is ironic that x-ray diffraction, which so readily highlights regular-ity and order, should encounter particular difficulties in distinguish-ing the symmetry which, perhaps more than any other form of repetition, has intrigued artists, scientists, engineers and architects throughout history. This is the relationship between an object and its mirror image. The x-ray diffraction pattern recorded from a right-handed helical molecule like that in figure 3 would, under normal circumstances, be identical to that recorded from the left-handed helix produced by reflecting it in a mirror. Despite these limitations, it was through the development of a new and elegant application of x-ray diffraction that the capability was developed for determining molecular handedness. This technique exploits the fact that an x-ray wavelength can be identified at which a particular atom will anoma-lously change the phase of the scattered x-ray. This change results in intensity differences which can be analysed to establish unequivo-cally the handedness of the molecule.

Characterising Hierarchies

Structure and function in the natural world is based on hierarchy. This can be seen in relationships between subatomic particles, atoms, molecules, macromolecules, macromolecular assemblies, organelles, cells, organs, individuals and communities. Within the size range from about 1 Å to 1000 Å x-ray diffraction has played a crucial role in revealing the details of this hierarchy. The subatomic particles are below the lower limit of this range and in molecular descriptions of matter atoms are regarded as the fundamental building blocks. Although the configuration of the electrons of an atom are perturbed during the formation of a chemical bond, this is limited predominantly to the outer electrons. To a very large degree the atom retains characteristics of its identity when it becomes part of a molecule, e.g., a well defined van der Waals radius, which determines its overall size, and a covalent radius, which can be used in calculating covalent bond lengths. The accepted values for these parameters, and others such as ionic radii, lengths of hydrogen bonds and values for angles between covalent bonds, are based on structures determined by x-ray diffraction.

Molecular Structures

The macromolecules on which the structures and biochemical capabilities of organisms are based are predominantly linear polymers constructed from a relatively small number of distinct monomer building blocks, e.g., nucleic acids from four different nucleotides and proteins from twenty different amino acids. Following polymerisation, not only do the atoms within the monomers making up the polymer chain retain many of their characteristics but also the distinctive characters of individual monomers are retained. This is well illustrated by the structures and properties of proteins where amino acids are commonly divided into categories according to the character they retain within the proteins. For example, phenylalanine, alanine and valine are categorised as hydrophobic and histidine, lysine and arginine as basic.

    The assumption that structural characteristics of a unit, such as an amino acid, are largely preserved when the unit is incorporated into larger and more complex structures was at the heart of strategies developed in the late 1940s and 1950s for elucidating the structures of important biological macromolecules. In particular, this assumption was essential to Pauling's highly coordinated approach for the determination of protein structures[9]. With the benefit of detailed stereochemical information on the structures of individual amino acids and on the peptide group, which links successive amino acids along a protein chain, it was possible to construct molecular models for proteins which allowed free rotation about single bonds. This enabled the overall conformation of the chain to be adjusted to achieve the best fit between the x-ray diffraction calculated for the model and that observed from the fibrous proteins keratin and silk[10,11]. Such studies led to the discovery of the α-helix (figure 4)

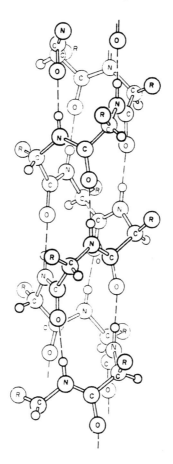

Figure 4. Right-handed α-helical conformation of a protein chain.

and extended β-sheet structures. These have dominated thinking during the succeeding forty years not only about fibrous proteins but also about globular proteins, such as enzymes, which are central to metabolic processes and in which the protein chain is typically coiled irregularly into a globule.

The amino acid sequence along a protein chain is the primary structure of the protein and the folding of the chain into structures like the α-helix and extended β-sheet is the secondary structure. However, while globular proteins typically contain regions of the protein chain with conformations similar to the α-helix and extended β-sheet structures, the overall conformation, called the tertiary structure, is typically much less regular with the superficial appearance of a badly wound ball of string. Despite this appearance it is important to emphasise that chain folding within a globular protein is highly specific. It represents a minimum energy conformation, which is regained following thermal denaturation if a solution of the denatured protein is cooled under appropriate conditions. Despite this understanding, progress in applying the laws of physics and chemistry to predict the tertiary structure of a protein from its primary structure has been rather limited; our knowledge of the detailed structure of globular proteins is still dominated by results from x-ray diffraction.

X-ray protein crystallography has its origins in the 1930s in the work of Bernal and Crowfoot (later Hodgkin)[12] who made the important observation that to maintain order within a protein crystal it was necessary to keep it in contact with the mother liquor from which it had been crystallised. The x-ray diffraction patterns recorded from these crystals indicated the scale of the problems which would need to be overcome in a successful structure determination; it was hardly surprising that the first globular protein structure was not solved until almost thirty years later. This was the product of a powerful and sustained programme by Perutz and Kendrew and their co-workers[8,13] on proteins whose function was the transport and storage of oxygen. Of these the first to be determined was myoglobin, an oxygen storage protein with a molecular weight of about 17000. The first Fourier synthesis to be described was at the relatively low resolution of about 6 Å (figure 5). The structure

Figure 5. Drawing of a model of the myoglobin molecule at 6 Å resolution (from reference 12). The hæm group is shown as a foreshortened dark grey disc. Each of the "sausage like" structures corresponds to a region of α-helix. Further refinement of this structure showed that the hæm group orientation illustrated in this model should be modified by a clockwise rotation of approximately 40° about a horizontal axis in the plane of the disc.

revealed in this synthesis consisted almost entirely of tubes of electron density which were identified, with some certainty, as α-helices. This assignment was confirmed in subsequent Fourier syntheses at about $2\,\text{Å}$[13]. Further, by exploiting the anomalous scattering from the iron atom in the hæm group, it was shown that all the α-helices were right-handed. In addition to its importance in solving the first structure of a globular protein, the analysis of myoglobin was important because of its relationship to hæmoglobin, in which each of the four symmetrically arranged subunits is similar to a complete myoglobin molecule (figure 6). The arrangement of subunits within a complex molecule like hæmoglobin is designated as quaternary structure. The biological exploitation of the capabilities stemming from quaternary structure, in which identical subunits are arranged symmetrically, was discussed in the section "Functional Advantages of Symmetry in Biological Systems".

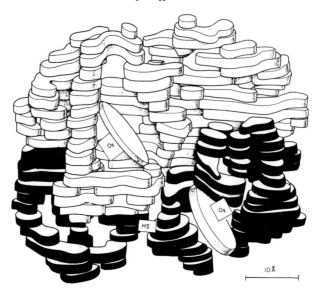

Figure 6. Model showing the four subunits in hæmoglobin (from reference 8). Two subunits are shown black and two white—there is a hæm group in each subunit but only two can be seen. This structure and those of a large number of mutants have subsequently been refined to atomic resolution.

Macromolecular Assemblies

Muscle provides one of the most extensive and coherent examples of structural hierarchy in biology—figure 7 shows that this structural coherence extends to spacings of many micrometres. Thus, for an understanding of muscle function, it is necessary to bridge the gap between the information on molecular structure and organisation obtained from light microscopy and that from a combination of x-ray diffraction and electron microscopy. Muscle is a complex assembly of proteins embracing the metabolic capabilities characteristic of globular proteins and the structural capabilities of fibrous proteins, as might be anticipated for a system which converts chemical energy into mechanical energy and is able to sustain tension. Within this assembly individual actin and myosin molecules, which respectively make up the thin and thick filaments, perform various functions. The understanding of these is aided by information obtained from single crystal x-ray diffraction analyses of the three-dimensional structures of the extracted molecules.

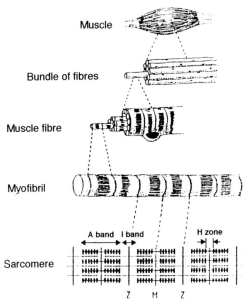

Figure 7. Representation of the structural hierarchy in muscle (from reference 13). In the bottom section of the figure the thin actin filaments are shown interdigitated between the thick myosin filaments.

X-ray fibre diffraction studies of intact muscle have yielded structural information which provides a framework within which the more detailed information from single crystal structure determinations can be correlated. It is clear, from the principle of reciprocity in the relationship between an object and its diffraction pattern, that to observe structural features between about 1000 Å, the normal upper limit which can be studied by x-ray diffraction, and 10 000 Å or more, as seen by light microscopy, will require x-ray diffraction data to be recorded at very small angles. Joan Bordas and co-workers[14] have achieved this by exploiting the highly collimated beam of the Daresbury Synchrotron Radiation Source and sophisticated beam-line optics. They characterised structural periodicities in muscle of about 20 000 Å from diffraction data recorded at angles as small as 0.005° relative to the incident x-ray beam direction. The application of x-ray diffraction in characterising changes in molecular structure and organisation during muscular contraction is described in the section "Following Change".

Structure in Context

The determination by x-ray diffraction, in the 1950s and 1960s, of the structures of several biological macromolecules was widely recognised as a major technical achievement. Nevertheless, there remained dissident voices who questioned whether these structures were biologically relevant. These doubts were gradually set to rest as, almost without exception, each structure which was determined contributed, convincingly and often unexpectedly, to an understanding of how the molecule functioned. This is well illustrated by the understanding of the relationship between DNA structure and function as well as by the visualisation, from the work of Phillips and colleagues[15], of the interaction of the enzyme lysozyme with its substrate. This was the first enzyme structure to be visualised at atomic resolution and the subsequent thirty years have seen an ever increasing expansion of the number and complexity of protein structures which have been determined.

While biological macromolecules in general, and enzymes in particular, have well defined structures which persist in a wide range of environments, the ability to make well defined conformational changes, following specific interactions with other macromolecules, is also central to biological function. An early recognition of this was found in models for the enzymatic function of lysozyme, which proposed destabilisation of a covalent bond in the substrate through stereochemical distortion following binding in a cleft on the enzyme. Such distortions are of common occurrence and a variety of effects associated with specific binding of substrates have been investigated by x-ray diffraction analyses of single crystals. Of particular interest are subtle effects such as binding of an effector molecule, at a site on an enzyme remote from the active site, which induces a change in the stereochemistry at the active site with the effect of turning enzyme activity on or off.

Relationships Between Structure and Function

The elucidation of the stereochemical basis for specificity in the immune response is one of the most challenging areas in structural molecular biology. The contribution which x-ray diffraction can make to this problem is well illustrated by the 'canyon hypothesis' for the origin of specificity in the binding of animal viruses such as the human immunodeficiency viruses (HIV) to cellular receptor sites. The immune system in animals is capable of rapid adaptation in response to new sources of infection and it is assumed that crucial to this capability is regular 'surveillance' of foreign bodies followed by the production of antibodies which bind specifically to the foreign body. For this defence mechanism to be effective it is crucial that the antibody should bind at a site on the foreign body which impedes its pathological activity. The 'canyon hypothesis' has its origins in detailed scrutiny of the structures determined by x-ray diffraction for a series of spherical viruses. The principle of constructing these viruses according to an icosahedral arrangement of identical sub-units was discussed previously. The symmetry in viruses of this type has been exploited in the determination of their structure by x-ray analyses of virus single crystals[16,17]. Some perspective on the significance of this structure determination for the technical development of x-ray diffraction is provided by the fact that the complete protein coat has a molecular weight of many million and the resolution of the analysis is sufficient for individual atoms to be distinguished. The folding of the protein chains and the overall icosahedral symmetry of foot and mouth disease is illustrated in figure 8. It is this degree of detail which has allowed hypotheses to be formulated about the atomic interactions which confer specificity on the binding of a virus particle to a cellular receptor.

The 'canyon hypothesis'[18] was based on the structure of human rhinovirus and the existence of a deep surface depression, or canyon, encircling each of the twelve fivefold symmetry axes specific to an icosahedral structure. The central thrust of the hypothesis is that the atomic groupings, whose three-dimensional structure confers specificity in virus binding to the host cell receptor, lie deep within the canyon. It is further proposed that the host cell is prevented from developing an immune response, which would specifically block the receptor binding site inside the canyon, because the canyon is too small to allow surveillance of the site through antibody penetration. This hypothesis is valuable since it can be tested by observing the effect of specific stereochemical changes, inside the canyon and around its edges, on virus infectivity. Developments in

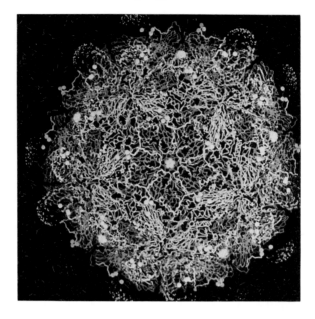

Figure 8. Foot and mouth disease virus showing the folding of the protein chain (from reference 16).

biotechnology allow such changes in virus structure to be made and the 'canyon hypothesis' to be established as representing one strategy which has been exploited by viruses in their continuing battle against host cell immunological responses. It is important to emphasise that not only does the determination of a structure as large and complex as a spherical virus represent a formidable analytical achievement but also the collection of the necessary x-ray diffraction data only became possible through major developments in x-ray detectors and sources. Not the least of these developments was the construction of x-ray synchrotron radiation beamlines specifically designed for investigating molecular structure and organisation.

Specificity in Partially Ordered Systems

Knowledge of the stereochemistry of the interactions between an enzyme and its substrate, or a virus and its receptor, is crucially important for programmes directed at the rational design of drugs. It is also important to recognise that the total drug delivery process will typically involve interactions with a wide variety of systems. For many of these systems, such as the membranes which envelop and partition a cell, the functionally important state is one of partial order whose characterisation, like that of the structure and organisation of polymer molecules in fibrous materials, has depended on techniques for data collection and analysis beyond those employed in single crystal analyses. As in the previous discussion of muscle it is important to recognise that while the techniques which have been developed for studying partially ordered systems (such as solutions, sheets and fibres) can yield unique information this will typically be less detailed than that obtained in studies of single crystals. It is important, therefore, that wherever possible the information obtained on these systems should be augmented by information from single crystal studies of their individual components.

Important aspects of the characterisation of partially ordered systems are the determination of the distance over which a particular structural feature persists and the nature of the structural relationships which exist between neighbouring regions. The availability of the high brilliance microfocus beam at the recently commissioned European Synchrotron Radiation Source (ESRF) offers the possibility for investigating such structural variation with a spatial resolution of about 1 μm. The power of this facility is well illustrated by recent studies by Mahendrasingam and colleagues[19] on spherulites of the organic polymer Biopol. In these studies a computer controlled stage allowed the spherulite to be tracked in two dimensions perpendicular to the x-ray beam in steps as small as 1 μm. An array of diffraction patterns recorded using this facility is illustrated in figure 9, which vividly demonstrates that on a spatial resolution of 10 μm the crystal lamellæ are systematically oriented with an axis along a radius of the spherulite. Quite apart from the understanding these experiments have given about the processes of nucleation and growth in spherulites, the techniques developed in them can be expected to have general applicability in the investigation of texture in materials and structural variation in biological materials. Examples include investigations of crystallite type, size and orientation in the mineralisation of the organic matrix in bone and the shells of molluscs.

Figure 9. An array of x-ray fibre diffraction patterns recorded from the central region of a spherulite of biopol. Each pattern was recorded from an area ~10 μm in diameter as the specimen was tracked systematically across the microfocus x-ray beam. Each diffraction pattern was recorded with an exposure time of 1 second and the array of patterns provides a map of the variation of polymer crystallinity and orientation across the spherulite.

Following Change

The low intensity of diffracted x-rays, compared to that of the incident beam, was emphasised in the introduction and is reflected in the fact that even with the commercial x-ray generators available in the 1950s and 1960s a typical x-ray diffraction experiment would involve collecting data over a period of many hours or even days. Indeed a very fine and much reproduced diffraction pattern from a square array of four fibres of DNA in the A conformation, each 80 μm in diameter, was recorded over a period of 10 weeks[6].

With such long exposure times it might have been concluded that x-ray diffraction would never have the capability of following the structural transitions important in biological function. For more than three decades the contractile cycle in muscle has been recognised as presenting the most formidable challenge in the development of dynamic descriptions of centrally important biological phenomena. This challenge has been all the more tantalising since it was over forty years ago that the static view of the structure of muscle, developed largely from electron microscope studies, suggested in general terms how muscular contraction might occur[20,21]. In this so called sliding filament model a regular array of thin actin filaments slides between a regular array of thick myosin filaments. It stimulated an enormous variety of biochemical and biophysical investigations aimed at identifying the structural features, on the actin and myosin filaments, which were responsible for drawing one family of filaments past the other. These studies were aided by the identification of conditions which allowed a dissected muscle to be maintained in a physiologically active state while it was studied by x-ray diffraction. In particular, the ability to take a muscle repeatedly through the contractile cycle was exploited by synchronising the opening and closing of an x-ray shutter with the state of contraction of the muscle.

These experiments received an enormous boost by the availability of synchrotron radiation sources whose brilliance was many orders of magnitude greater than that of the most powerful rotating anode x-ray generators. Indeed, the first attempts to use an electron accelerator designed for high energy physics in x-ray diffraction studies used a sample of muscle[22]. The impact of specifically designed synchrotron sources on x-ray diffraction is well illustrated by the studies of Huxley and co-workers[23]. The analysis of such x-ray diffraction data, and its correlation with a vast amount of data from electron microscope, biochemical and mechanical studies, has been predominantly in terms of the 'swinging crossbridges' model. These crossbridges originate on the myosin filaments and, at the critical state in contraction, are exposed as a consequence of a series of stereochemical events prompted by the arrival of the nerve impulse. The exposed headgroups on the crossbridges bind to sites on the actin filaments which, through a coordinated change in the inclination of the crossbridges, are translated with respect to the myosin matrix.

Perhaps the most ambitious and comprehensive attempt to interpret x-ray diffraction data in terms of the swinging crossbridge model is by Squire and co-workers[24]. Their analysis used diffraction data from muscles of bony fish recorded at the Daresbury Synchrotron Radiation Source. This fish muscle is particularly suited to diffraction studies because it is more highly ordered than other vertebrate muscles. Data was collected for a period of about 200 ms while the tension in the originally relaxed muscle gradually increased to its maximum value. The time required to record each diffraction pattern was 5 ms, giving data on about 40 intermediate stages during the contraction process and providing the raw material for a film, appropriately entitled "Muscle — The Movie"[24].

Examples have previously been described of structural changes in globular proteins, which in the case of hæmoglobin were associated with binding and release of oxygen and in the case of lysozome with cleavage of a specific chemical bond in a substrate. The structural information obtained from these studies, and from the large number of similar investigations of enzymes which they stimulated, have been extremely productive in terms of identifying the stereochemical requirements for a great variety of biochemical processes. However, the prospect of a full dynamic characterisation remains tantalisingly elusive. There are two crucial obstacles to be overcome in the pursuit of dynamic descriptions by x-ray diffraction. First, the very great speed with which enzyme reactions typically proceed and, second, the difficulty of ensuring that the reaction is proceeding synchronously across a crystalline specimen which contains millions of millions of molecules. An early attack on this problem by Wyckoff et al[25] exploited the availability of a poor substrate which slowed down the action of the enzyme ribonuclease. X-ray diffraction studies of enzymes functioning normally became credible with the advent of synchrotron radiation sources which allowed diffraction data to be collected with exposure times of milliseconds[26].

While some x-ray dynamical studies of enzyme action have used monochromatic beams of x-rays, particular efforts have been directed at increasing the incident intensity by using a wide range of wavelengths selected from the continuous distribution of synchrotron radiation. Although radiation from a synchrotron source has a continuous wavelength distribution, it is not continuous in time. A typical synchrotron beam consists of pulses each about 100 ps long and the collection of a data set within the timescale of a single pulse has become a compelling goal. Even if the brilliance of a synchrotron x-ray beam is adequate for data collection on this timescale, formidable problems remain to be solved[27,28] before a truly dynamic picture of structural changes during enzyme/substrate interaction can be obtained. Techniques using an incident x-ray beam with a continuous wavelength distribution are currently at the forefront of innovation in x-ray diffraction. Their designation as Laue techniques recognises their basic similarity to the first x-ray diffraction experiments, conducted over eighty years ago, and it is therefore appropriate to conclude this review of experimental and analytical innovation in x-ray diffraction with this topic.

Modelling and Design

There is hardly an area of molecular biology or materials science which has not been illuminated by the application of x-ray diffraction. The enormous range, not only in the nature of the structures studied but also in the type of information obtained, bears witness to the very great diversity which is possible in the employment of x-ray diffraction techniques. The economic theories expounded by Karl Marx may be out of fashion but this has not prevented his rallying cry that "The philosophers have only interpreted the world in various ways; the point is to change it" becoming the dominant ethos of developed countries in the last decade of the twentieth century. It is the intellectual descendents of the Natural Philosophers, and in particular physicists and chemists, who have contributed so much to understanding and interpreting both the animate and inanimate world and who have modelled it so convincingly in terms of atoms represented by coloured balls. No one involved in this process can fail to be awed by the elegance and effectiveness of the solutions found in nature for optimising three-dimensional structure to meet functional needs.

By comparison to the syntheses which take place routinely in living systems, the solutions developed in technologies directed at responding to the call to change the world by the synthesis of new molecules have been very modest. The advances in biotechnology over the last decade have dramatically extended these capabilities, with x-ray diffraction and, particularly, protein crystallography being looked to for specific guidance on the chemical changes required to produce a more effective drug or some desirable new property in a material. The possibilities for rational design of drugs and materials are truly enormous and it would be irresponsible for technologically advanced societies to turn their backs on these opportunities. However, enthusiasm for the benefits, especially when driven by a too selfish approach to wealth creation, should not be allowed to override caution and indeed a little humility. As the images on the computer graphics screens become ever more impressive we need to become more, not less, self critical of the validity of our models and the dangers of too arrogant a faith in them.

Acknowledgements

I am grateful to Caroline Tahourdin, Helen Moors and Mike Daniels for help with the preparation of this review.

References

(1) M von Laue "Nobel Lecture in Physics 1915" *Nobel Lectures Physics 1901–22* Elsevier: Amsterdam p 347 (1967)

(2) W L Bragg "Nobel Lecture in Physics 1916" (lecture delivered 1922) *Nobel Lectures Physics 1901–22* Elsevier: Amsterdam p 370 (1967)

(3) J D Watson & F H C Crick "A Structure for Deoxyribose Nucleic Acid" *Nature* **171** 737 (1953)

(4) M H F Wilkins, A R Stokes & H R Wilson "Molecular Structure of Deoxypentose Nucleic Acid" *Nature* **171** 738 (1953)

(5) R E Franklin & R G Gosling "Molecular Configuration of Sodium Thymonucleate" *Nature* **171** 740 (1953)

(6) M H F Wilkins "Nobel Lecture in Physiology or Medicine 1962" *Nobel Lectures Physiology or Medicine 1942–62* Elsevier: Amsterdam p 754 (1964)

(7) F H C Crick & J D Watson "Structure of Small Viruses" *Nature* **177** 473 (1956)

(8) M F Perutz "Nobel Lecture in Chemistry 1962" *Nobel Lectures Chemistry 1942–62* Elsevier: Amsterdam p 653 (1964)

(9) L Pauling "Nobel Lecture in Chemistry 1954" *Nobel Lectures Chemistry 1942–62* Elsevier: Amsterdam p 429 (1964)

(10) L Pauling, R B Corey & H R Bransom "The Structure of Proteins: Two Hydrogen-Bonded Helical Configurations of the Polypeptide Chain" *Proc. Natl. Acad. Sci. U.S.* **37** 205 (1951)

(11) L Pauling & R B Corey "Configurations of Polypeptide Chains with Favoured Orientations Around Single Bonds: Two New Pleated Sheets" *Proc. Natl. Acad. Sci. U.S.* **37** 729 (1951)

(12) J D Bernal & D Crowfoot "X-Ray Photographs of Crystalline Pepsin" *Nature* **133** 794 (1934)

(13) J C Kendrew "Nobel Lecture in Chemistry 1962" *Nobel Lectures Chemistry 1942–62* Elsevier: Amsterdam p 676 (1964)

(14) G P Diakun & J E Harries "Description of an Ultra Small Angle X-Ray Diffraction Camera for Studies of Striated Muscle" *Synchrotron Radiation and Biophysics* (ed S S Hasnain) Ellis Horwood: Chichester p 243 (1990)

(15) D C Phillips "The Three-Dimensional Structure of an Enzyme Molecule" *Sci. Am.* **215** 78 (1966)

(16) M G Rossmann *et al* "Structure of a Human Common Cold Virus and Functional Relationship to other Picornaviruses" *Nature* **317** 145 (1985)

(17) R Acharya *et al* "The 3-Dimensional Structure of Foot and Mouth Disease Virus at 2.9 Å Resolution" *Nature* **337** 709 (1989)

(18) MG Rossmann "The Canyon Hypothesis—Hiding the Host Cell Receptor Attachment Site on a Viral Surface from Immune Surveillance" *J. Biol. Chem.* **264** 14587 (1989)

(19) A Mahendrasingam *et al* "Microfocus X-Ray Diffraction of Spherulites of Poly-3-hydroxybutyrate" *J Synchrotron Radiation* (1995) in press

(20) A F Huxley & R Niedergerke "Interference Microscopy of Living Muscle Fibres" *Nature* **173** 971 (1954)

(21) H E Huxley & J Hanson "Changes in Cross Striations of Muscle During Contraction and Stretch and their Structure Interpretation" *Nature* **173** 973 (1954)

(22) G Rosenbaum, K C Holmes & J Witz "Synchrotron Radiation as a Source of X-Ray Diffraction" *Nature* **230** 434 (1971)

(23) H E Huxley *et al* "Changes in the X-Ray Reflections from Contracting Muscle During Rapid Mechanical Transients and their Structural Implications" *J Mol. Biol.* **169** 469 (1983)

(24) J M Squire, J Harford & E Morris "Muscle—the Movie" *Image Processing* p 22 (Spring 1993)

(25) H W Wyckoff *et al* "Design of a Diffractometer Flow Cell System for X-Ray Analysis of Crystalline Proteins with Applications to the Crystal Chemistry of Ribonuclease-S" *J Mol. Biol.* **27** 563 (1967)

(26) J Hajdu *et al* "Millisecond X-Ray Diffraction and the 1st Electron Density Map from Laue Photographs of a Protein Crystal" *Nature* **329** 178 (1987)

(27) K Moffat *et al* "Time Resolved Macromolecular Crystallography: Principles, Problems and Practice" *Phil. Trans. Roy. Soc. Lond.* **A 340** 175 (1962)

(28) G A Petsko "Art is Long and Time is Fleeting: the Current Problems and Fututre Prospects for Time Resolved Enzyme Crystallography" *Phil. Trans. Roy. Soc. Lond.* **A 340** 323 (1962)

Synchrotron Radiation

Ian Munro
Daresbury Laboratory

V ery shortly after the pioneering work of Röntgen in 1895 Liénard (1898) and Schott (1912) effectively laid the foundations for the development of the classical, electrodynamical theory of synchrotron radiation. This early work followed the discovery of the electron and was greatly in advance of the development of particle accelerators which commenced in the 1920s and grew slowly until the beginning of the Second World War. During the 1940s and 1950s physicists produced high energy charged particles using synchronous accelerators which were large machines where the accelerating charged particles were constrained by magnetic fields to move in circular orbits. The energies needed for elementary particle physics experiments require charged particles to be accelerated to close to light speed before extraction for collision experiments. It was always recognised that the primary cause of energy loss during the acceleration of charged particles to high energies was the emission of electromagnetic radiation. In 1944 Ivanenko and Pomeranchuk in Russia and slightly later Schwinger (1946) in the USA and Sokolov and Ternov (1948) in Russia calculated the fundamental properties of this so called synchrotron radiation.

By the end of the 1940s the first direct observations of synchrotron radiation, during the acceleration of electrons, had already been made by Blewett, Elder and others. This represented the beginning of the synchrotron radiation era which has led to such an immense range of scientific endeavour in most of the major countries of the world. During the 1950s several experimental measurements were made, to compare with the theoretical predictions, in Moscow (the Lebedev Institute) and the USA (National Bureau of Standards). The

X-RAYS: The First Hundred Years
Edited by Alan Michette and Sławka Pfauntsch © 1996 John Wiley & Sons Ltd

principal contribution was probably that of Tomboulian and colleagues at the Cornell 300 MeV electron synchrotron in the USA.

The specific property of synchrotron radiation which initially caught the attention of scientists from outside high energy physics was the smooth continuous emission spectrum (figure 1). This emission is calculable exactly in terms of direction, flux and polarisation and it extends from the far infrared to x-rays (figure 2). The 'white' continuum, particularly in the x-ray region, provided a unique opportunity to develop fundamental atomic and molecular spectroscopy in a photon energy regime where alternative sources either did not exist or were exceptionally difficult to maintain and use. The construction of electron accelerators to support the large scale of activity in high energy particle physics throughout the 1960s offered the possibility to carry out such pioneering experiments in the applications of synchrotron radiation. Although these experiments were secondary to those in high energy physics, important advances were made in Italy (Frascati), the USA (National Bureau of Standards and Wisconsin), Germany (Hamburg), Japan (Tokyo), USSR (Moscow and Yerevan) and the United Kingdom (Daresbury and Glasgow).

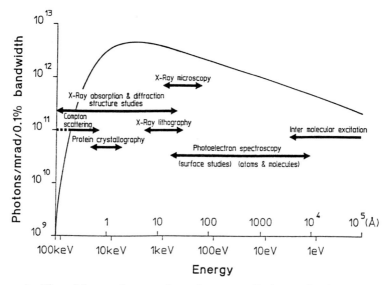

Figure 1. The white continuum of synchrotron radiation emitted from the 5 T wiggler magnet at the SRS Daresbury Laboratory. The continuum extends to a frequency of 500 MHz (equivalent to about 10^{-5} eV).

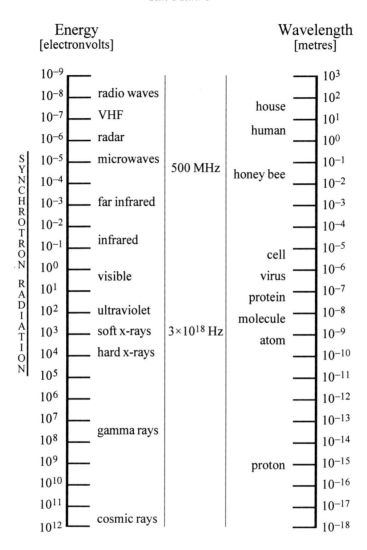

Figure 2. Synchrotron radiation in the electromagnetic spectrum.

In all this early work the use of synchrotron radiation was severely hampered by the fundamental property of a synchrotron, namely that the energy of the accelerated particles increases with time until they are extracted for high energy physics experiments. This means that the emission spectrum of synchrotron radiation from synchrotrons is strongly time dependent. What was needed was a machine in which the energy of the electrons remained constant over a long time, which would offer the potential synchrotron radiation user the advantages of fixed spectrum, photon source stability, time stability and minimal radiation hazards. The first experiments with storage rings (beam colliders) soon stimulated the parasitic exploitation of their 'unwanted' electromagnetic radiation by the 'first generation' of synchrotron radiation users—a truly symbiotic relationship with high energy physics users!

During this first era of synchrotron radiation exploitation, in the 1970s, there were many fundamental and far reaching new developments. The first high resolution x-ray spectroscopy experiments, the recognition of the unique brightness of synchrotron radiation for studies of small crystals and for small angle scattering, microscopy and imaging experiments in the x-ray region, all laid the foundation for the tremendous advances in x-ray science which have taken place in the subsequent twenty years.

The considerable inconvenience of working parasitically alongside high energy physics users and the rapidly expanding scale of scientific activity using synchrotron radiation led to the design, construction and operation of storage rings whose sole purpose was to provide radiation for diffraction, scattering and spectroscopy. The first storage ring wholly dedicated to synchrotron radiation research was Tantalus built in 1968 in the USA (Wisconsin). By the start of the 1980s several such sources were under construction and the first purpose built, dedicated x-ray source for synchrotron radiation usage was completed and operational at Daresbury in the UK by 1981.

Throughout the 1970s experiments using synchrotron radiation were almost exclusively performed with radiation emitted from dipole bending magnets. However, it was soon recognised that the use of multiple dipoles should greatly enhance the flux, brightness and brilliance of the photon beams generated from such devices, and a small experiment, using a wavelength shifting wiggler, was conducted in Wisconsin early in the decade.

Therefore, within the span of the 1970s, the area of accelerator physics began to be converted from one in which the minimisation of synchrotron radiation was of great importance to one in which the maximisation and optimisation of synchrotron radiation was the sole purpose for the construction of the accelerator!

Fundamental Properties of Synchrotron Radiation

For a synchrotron radiation user the properties of the source are defined in terms of the combination of two groups of parameters. First, the radiated power spectrum, degree of polarisation and angular distribution of the radiation from a single electron is calculable exactly using formulæ derived from the work of Schwinger and others. Second, the behaviour of the photon source is given by combining the calculations for an individual electron with the collective behaviour, which can also be calculated, of all the electrons in the storage ring beam—the number of electrons involved is typically around 10^{12}. At any point in the ring the electron beam has a known spatial distribution and a calculable emittance, the product of the beam size and its angular divergence. In early storage rings the emittance was approximately 10^{-6} metres radians (m rad) whilst for the state of the art devices of the 1990s the target value is about 10^{-9} m rad. For a given flux (the number of photons per second in a particular wavelength range) a lower emittance leads to a higher brilliance (the flux per unit area of the source per unit solid angle).

In an electron storage ring (figure 3) apart from electron scattering by residual gas atoms in the beam chamber the only mechanism

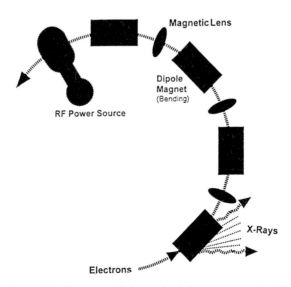

Figure 3. A schematic diagram of the principle components of an electron storage ring used as a synchrotron radiation source.

for energy loss is through synchrotron radiation so that nearly all the input power appears, in principal at least, as usable radiation. This makes synchrotron radiation sources extremely efficient. In addition, since the electrons must be accelerated in an ultrahigh vacuum (about 10^{-9} mbar) the source properties, i.e., the shape and position of the electron beam, are not modified in any way by the surroundings. The radiation from each electron perturbs its orbit by a very small amount and therefore immense amounts of power (up to one megawatt of x-rays) may be radiated from a storage ring without major changes to the beam. X-rays which are not extracted from the ring are lost by absorption or scattering in the interior wall of the orbit chamber. The resulting heat load must be efficiently dissipated, normally by the use of a complex water cooling system.

The total radiated power from a storage ring is given by $P=4\times10^{-3}E^3BI$ W per horizontal milliradian, where E is the electron energy in GeV, B is the dipole field in tesla and I is the electron beam current in amps. This illustrates the strong dependence of the emitted power on the electron beam energy — increasing the energy from 1 GeV to 8 GeV will produce around 500 times more power, other things being equal. The total power also scales with the magnetic field strength and the current. The Synchrotron Radiation Source (SRS) at the Daresbury Laboratory operating at 2 GeV and at about 250 mA yields approximately 10 W per horizontal milliradian from a dipole magnet.

The unique white continuum associated with synchrotron radiation sources is a consequence of the extremely high electron speed with respect to a stationary observer. The emission is strongly peaked in the forward direction, just like a stone thrown from a slingshot, and so the observer sees radiation only from a very small opening angle, typically less than 1 mrad, corresponding to a very short length of the electron orbit (see figure 4). The radiation pulse is produced for a very short time which will appear to the observer to be further shortened, due to relativistic effects, to about 10^{-19} s. The frequency analysis of such a short pulse reveals very high harmonics with angular frequencies up to about 10^{12} times the fundamental orbit frequency, which in most storage rings is between 10 MHz and 500 MHz. In practice the individual harmonics cannot be resolved

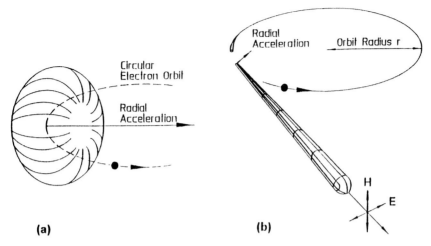

Figure 4. Comparison between the radiation from a) a nonrelativistic charged particle and b) one travelling close to the speed of light.

and the radiation appears to have a continuous spectrum extending from the fundamental, which in the case of the SRS at Daresbury is 500 MHz corresponding to a wavelength of 60 cm, to the shortest wavelengths of about 0.1 nm (see figure 1).

A useful quantity for comparing synchrotron sources is the critical wavelength λ_c (or corresponding energy) which divides the power spectrum in half, i.e., half the total power is radiated at wavelengths shorter than λ_c and half at longer wavelengths.

The Evolution of 'Photon Factories'

The properties of synchrotron radiation can be exploited at any point on the beam orbit where the electrons are being accelerated by the magnetic field. For technical convenience, individual beam ports usually accept up to 50 mrads of arc and each beam port may be able to serve anything from one to nine experiments simultaneously by dividing the photon beam into segments. In the great majority of cases there are as many beam ports and experimental stations located around the storage ring as either cost or space will allow, leading to veritable 'factories' for photon experiments. As an example, the Daresbury SRS has approximately 40 stations around its perimeter (figure 5) which is close to the maximum number feasible within the

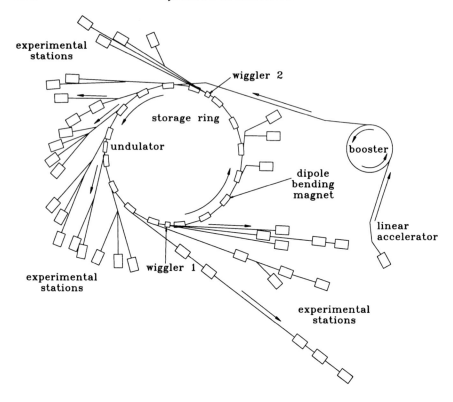

Figure 5. Schematic diagram of the SRS at Daresbury. The storage ring has a diameter of 30 m.

space available. Thus, although the initial cost of a storage ring may be large (anything from several million to several hundred million pounds), it is a very cost effective approach to the exploitation of high brilliance x-ray beams.

During the 1980s major developments took place in the technologies required to deliver the x-ray beam to the sample in a totally controlled manner. The principal component for this is a beamline leading tangentially from the storage ring perimeter and maintained usually at ultrahigh vacuum. Each beamline contains a device for reducing the spectral range of the radiation, either a grating or crystal monochromator, to provide a single wavelength or a range of wavelengths and the spectral resolution, usually about 10^{-3} or better, required for a given experiment. Each beamline also contains one or more mirrors which guide and focus the beam onto the sample.

The type of science undertaken at storage rings is primarily 'small' science carried out by relatively small groups of experimenters. This has proved to be of great benefit to the synchrotron radiation community because the individual skills from many different scientific and technical areas rapidly spread around each ring to the common good and advantage of all users. There have been striking advances in the design, construction and operation of x-ray monochromators, which are able, in some cases, to combine cooling (to minimise the heat load) with focusing and monochromatisation in a single crystal. Immense strides have been made in the enhancement and measurement of surface shape and surface texture of mirrors which are now capable of reflecting short wavelengths (of the order of 0.1 nm) with high efficiency. Developments in ultrathin coating techniques have led to the creation of multilayer devices, resembling synthetic crystals, by the deposition of multiple coatings with controlled layer thicknesses of a few tenths of a nanometre.

Beamlines may extend from a few metres up to a kilometre at the largest storage rings. The exceedingly high collimation of the beam has led to the need to control the source, i.e., the position and direction of the electron beam at the source points, to within very fine limits—typically a few micrometres. This is now achieved at many synchrotron radiation facilities using beamline photon monitors. These relay the photon beam position and direction at each beamline to the electron beam control system to allow uniform stability to be maintained simultaneously around many beamlines. In modern synchrotron radiation facilities, which have such careful control of the electron beam, the main causes of beam loss are probably due to failure of vacuum components such as bellows. When a leak is developed it may require the storage ring to be re-evacuated taking a substantial time (days or even weeks) to restore it to its working state.

Unlike the situation in high energy physics facilities, in the great majority of synchrotron radiation facilities radiation hazards are minimal since the stored beam is contained within a radiation enclosure. The radiation hazards associated with x-rays are constrained by the vacuum systems themselves, leading to a radiation free environment for the user community.

The Ultimate Prospects for Synchrotron Radiation Sources
During the 1980s the quest for high source brilliance and high flux at
the sample led to significant changes in the fundamental design of
synchrotron radiation sources. The flux of dipole radiation from
bending magnets (figure 6a) can be enhanced either by increasing
the horizontal beam aperture delivered to the sample or by increasing
the circulating current in the storage ring. However, a better result
can be obtained by increasing the number of source points which can

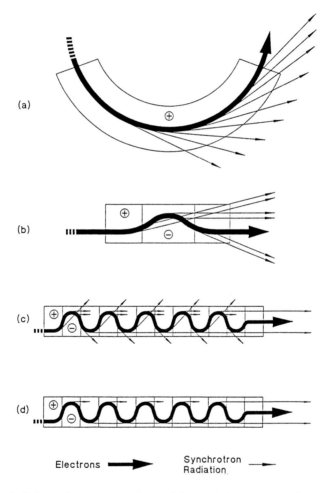

Figure 6. Schematic representations of the radiation patterns from   a) a dipole
with field about 1 T b) a high field wiggler (wavelength shifter) with field about
6 T c) a flux enhancing multipole wiggler with field about 2 T and d) a brilliance
enhancing undulator with field about 0.9 T.

be simultaneously focused on the sample and by reducing electron beam emittance and photon beam divergence. This can be achieved by using 'insertion devices' placed in a largely independent manner in the straight sections of storage rings between the dipole magnets.

Three basic types of insertion device have evolved, each of which has a significant benefit over a normal dipole magnet in terms of either spectral range, greater flux or higher brilliance. The simplest type is called a 'wavelength shifter' or 'three pole wiggler' (figure 6b). This device is simply a central dipole, usually operating at a high field strength, in association with two half field magnets to maintain the position and direction of the electron beam. The effect of a high field wavelength shifter is to reduce the value of λ_c. At the SRS at Daresbury wigglers with fields of 5 T and 6 T are used to provide high fluxes at approximately 0.1 nm for x-ray diffraction and scattering experiments, instead of operating at $\lambda_c \approx 0.4$ nm from the 1.2 T dipole magnets.

A further increase in output flux, important for many low efficiency experiments or for dilute samples, may be obtained using a second type of insertion device, the multipole wiggler (figure 6c). In this wiggler the magnetic field alternates in polarity along the length of the device and so it acts, in effect, as a series of independent dipole sources. A multipole wiggler with n poles produces n times the flux from a dipole magnet, radiating into about the same solid angle albeit from a much extended source. The construction of multipole wigglers with between 10 and 30 or more poles over lengths of up to 4 m has been made possible by the availability of suitable high field permanent magnet materials.

It is the third type of insertion device, the undulator (figure 6d), which has led to an evolutionary leap in brilliance over other x-ray sources. The gain in brilliance is based on a reduction in emittance and an increase in coherence of the photon source. An undulator is similar in construction to a multipole wiggler but with a reduced interdipole spacing. In this situation the electrons oscillate relativistically transverse to the electron beam direction and radiate at a wavelength determined by the interdipole spacing. The radiation is peaked in the forward direction and is quasi monochromatic. It is shifted due to the Doppler effect from λ_0, the magnet period in the undulator, to a series of spectral lines with wavelengths approximately $\lambda_0/2n\gamma^2$, where n is an integer and γ is the ratio of the electron's relativistic energy to its mass—typically a few thousand.

An example of the peaked quasi monochromatic radiation spectrum from an undulator, compared to those of a dipole magnet and a multipole wiggler, is shown in figure 7. The total power radiated by an undulator may be relatively small, less than 100 W, but it is concentrated into a beam of very small area. For the wavelength of the fundamental ($n=1$) the central brightness is exceptionally high, perhaps 10^6 times that of a bending magnet.

The total power radiated by a multipole wiggler can be very substantial and can create significant problems with the beamline optics, although its effect on the electron beam optics in the storage ring is usually not very important. The effect of a wavelength shifter or a high field superconducting dipole magnet on a typical storage

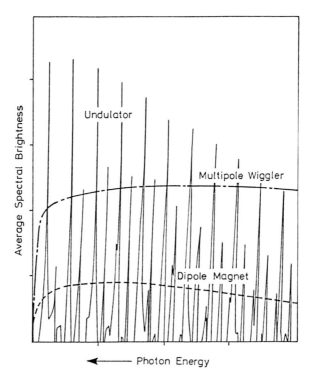

Figure 7. Spectral output from a dipole bending magnet, a multipole wiggler and an undulator. The brightness is plotted on a logarithmic scale.

ring lattice, however, is to produce an increase in the electron beam emittance, i.e., to increase the beam size, to the detriment of other users at all other points around the ring.

Of course, for all insertion devices the photon radiation pattern is governed not only by the photon beam divergence but also by the angular spread of the electron beam. For very long insertion devices (up to perhaps 10 or 20 m) where the magnet gap is small (this can be from a few millimetres to a few centimetres) the emittance of the storage ring must be extremely low in order not to lose the beam during its passage through the insertion device.

Low emittance storage rings are now being designed and constructed worldwide. In these the electron beam emittance is typically of the order of 10^{-9} m rad and in some cases this has lead to the creation of immensely large storage ring sources, with circumferences measured in kilometres. They are used to contain and accelerate electron beams with dimensions not much greater than that of a human hair. The beam positions are controlled to within a fraction of that width! The combination of small electron beam size and the reduced photon beam divergence of an undulator yields x-ray brilliances at least 10^6 times higher than those of bending magnet sources of the early 1980s, which themselves were considerably higher than those of sources available in a typical crystallography laboratory. This immense gain in brilliance is illustrated in figure 8.

By the late 1990s both linearly and circularly polarised radiation will be available from undulators with from 100 to, perhaps, 1000 periods. Photon source sizes of the order of 10^{-7} m^2, source divergences of 10^{-5} rad or below and intrinsic spectral purities of between 1% and 0.1% will yield unprecedentally high brilliances, exceeding 10^{20} photons s^{-1} mm^{-2} mrad^{-2} per 0.1% bandwidth.

Coherence

Coherence, which refers to the ability to create interference patterns when wavefronts from a source are recombined, is sometimes regarded as a property exclusive to lasers. However, undulators on modern low emittance storage rings are partially coherent sources. For small wavefront separations, smaller than the photon beam coherence length which is typically 1–10 μm, it has already been

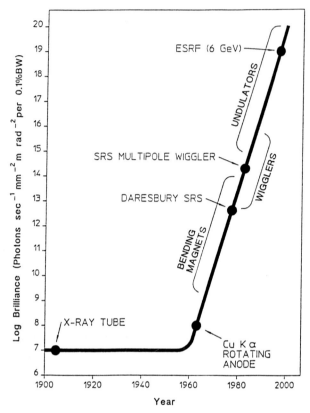

Figure 8. The evolution in x-ray brilliance during the past century.

possible to observe clear interference patterns. Soft x-ray interferometry and holography should soon become important applications of undulator radiation. Improved coherence, through better homogeneity of the magnetic field in long insertion devices, is a design aim of several future storage rings, such as Diamond at the Daresbury Laboratory. Current technology limits the magnet period to several centimetres with a minimum magnet gap of several millimetres but it may be possible in future to reduce the electron beam emittance and to develop micromagnet systems.

An alternative approach to improve coherence is to incorporate a free electron laser device within a continuously operating storage ring or in conjunction with an electron linear accelerator. A free electron laser is an undulator operating within an optical cavity. A convergence between synchrotron radiation and laser technologies towards the emergence of very large free electron laser systems represents the ultimate prospect for synchrotron radiation source development. An example of such a system is the recently proposed self-amplified spontaneous emission FEL, at DESY in Hamburg, which will be driven at 1 GeV by the linear accelerator of the superconducting test facility, Tesla, augmented by a low emittance injector and a longitudinal bunch compressor. This will lead to spectral brilliances in the range 10^{22} photons s^{-1} mm^{-2} $mrad^{-2}$ in 1% bandwidth—an immense, if not almost unthinkable, advance over the brilliances available from synchrotron radiation sources of only a few years ago (figure 9).

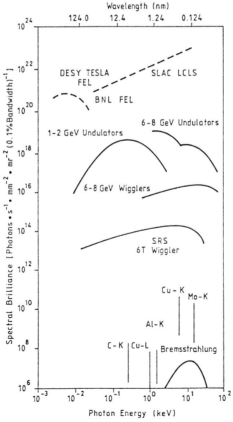

Figure 9. The ultimate limits of spectral brilliance envisaged by the present day synchrotron radiation community.

Why is Synchrotron Radiation Important?

An immense range of new x-ray technologies associated with the selection of photon energy and polarisation and with diffraction, scattering, spectroscopy, time resolved studies and imaging has evolved from the use of synchrotron radiation. The interaction between different technologies has led to a very substantial 'tunnelling' through the old 'potential barriers' between the disciplines of physics, chemistry and biology resulting in a multidisciplinary and multiple technique approach to the solutions of scientific problems. Throughout the 1970s and 1980s synchrotron radiation research probably gave a lead by this fresh approach to science which is now so essential to success in basic and applied research. There is hardly an area of science that has not benefited from synchrotron radiation.

The broadband nature of synchrotron radiation allows elements to be selected and their specific chemical environments to be explored. The radiation is sufficiently intense to allow state selection in gas molecules and to study clusters in condensed phase environments. Selecting its polarisation (linear or circular) enables directional effects in solids and surfaces and selection rules in atomic and molecular systems to be observed. The structures of molecules, surfaces and bulk materials can also be probed more effectively.

X-rays penetrate all forms of matter—solid, liquid and gas— and are scattered or absorbed to varying degrees. For almost a century they have provided one of the most versatile tools for the determination of atomic structures in solids, for chemical analysis and for the imaging of living systems. The evolution of synchrotron radiation sources based on insertion devices has now yielded the most intense sources of x-rays ever devised. Perhaps the most notable improvements brought about by these advances have been in x-ray diffraction methods, for example on very small crystals. Other techniques such as x-ray absorption spectroscopy and small angle x-ray scattering have become almost as routine as visible and ultraviolet spectroscopy.

Synchrotron radiation has led to a range of soft x-ray imaging, spectroscopy, diffraction and scattering methods, and has greatly enhanced our understanding of deep excitations in molecular states and the properties of molecules at surfaces and interfaces. The SRS at Daresbury, as a source designed and constructed specifically for x-ray research, has played a leading role in many of these new areas of science through its comprehensive research programme covering biological materials, surfaces, molecular sciences and engineering. Although all research carried out using synchrotron radiation is based at large centralised facilities, the work itself remains basically small scale, bench-top science where small groups from universities, industry and research institutions make use of experimental stations to carry out measurements in much the same way as they would have done in their own laboratory.

Biological Science

It is impossible in a short space to cover fully the scale and the quality of research which has been undertaken during the past thirty years. However, some examples from the research programme at the Daresbury Laboratory will serve to illustrate the benefits associated with the use of synchrotron radiation. Probably most notable in the last decade is the impact it has had on our ability to derive the structures of complex biological macromolecules and to correlate these with the ways in which they function. Three-dimensional structural information, obtained at synchrotron radiation sources around the world, underpins a great deal of modern molecular biology. Examples of such research, which have been undertaken at Daresbury, include determination of the foot and mouth disease virus structure, of the mechanism of muscle contraction and of a host of other large protein and enzyme structures. The crystallographic structural information obtained provides a template for the interpretation of the biochemistry and function of these systems.

The goal of much present day industrial research in biotechnology is the detailed investigation of biological structures to help to direct the design of new drugs. Synchrotron radiation provides a means for the collection of high quality, high resolution data in an extremely short time from large biological macromolecules where crystal sizes may be restricted to be less than about $100 \, \mu m^3$. The demand to pursue similar work is bound to increase during the next few decades in association with, for example, the human genome project which may well generate tens of thousands of new molecules many of which will require x-ray studies to establish the relationships between their structures and functions. New methods have emerged to help to solve the phase problem in macromolecular crystallography. Multiple wavelength anomalous diffraction, which has significant advantages over isomorphous replacement, is only available using a broad band x-ray source and has already been applied successfully to a study of small proteins.

Noncrystalline diffraction, including x-ray solution scattering, enables the conformation stages in macromolecules, including those in solutions and under physiological conditions, to be derived with much higher resolution than with methods which do not use synchrotron radiation. Despite the low scattering efficiency for x-rays, the high intensity of modern day synchrotron sources makes such experiments feasible. Future studies will be concerned with how macromolecules fold, since this has a crucial impact on the understanding of many biological problems associated with enzyme activity. This will, in turn, lead to protein structure prediction, which is of great importance in biotechnology. As the brilliance and flux from new synchrotron sources have grown, the ability to carry out time resolved studies in successively reduced timescales has become feasible. It is already possible to contemplate making time resolved diffraction and scattering measurements in less than one nanosecond! That is, before the material has had the chance to move or to reveal radiation damage effects.

The development of element selective x-ray absorption spectroscopy is a further method, complementary to macromolecular crystallography, which depends wholly on the use of synchrotron radiation for its successful application. In the x-ray absorption fine structure (XAFS) technique the measurement can be confined exclusively to the local environment of, for example, a metal site in a metalloprotein. This technique utilises the variations in absorption at energies just greater than absorption edges, illustrated schematically in figure 10. The details of these variations depend upon the local atomic environment. The advantage of x-ray absorption spectroscopy is that data can be collected on either noncrystalline (e.g., amorphous, glassy, liquid) or crystalline material with much the same resolution. It is therefore possible to obtain a penetrating

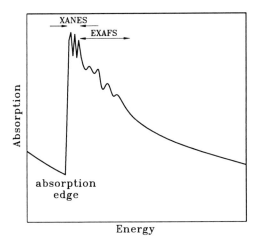

Figure 10. A schematic illustration of the energy ranges in which XANES (x-ray absorption near edge structure) and EXAFS (x-ray absorption fine structure) are carried out. For the copper K edge at 9 keV the structure extends for about 400 eV.

insight into local chemical, distance and coordination changes of site during a biochemical or biological reaction.

Materials and Surface Sciences

The area which has, and will continue to have, probably the most rapid development is studies of the condensed phase. Studies of the structural properties and behaviours of materials, interpreted in terms of the chemistry and physics of the solid state, have become an important field of research now that the focus of world attention has shifted from the basic sciences to the applied and strategic sciences.

The entire portfolio of methods available with synchrotron radiation has been applied separately and simultaneously to studies of subjects as diverse as electronic materials, polymers, composites, ceramics, magnetic materials and superconductors. As a result of improvements to the technology of detectors and data acquisition, real time *in situ* studies of industrially significant materials processing methods, such as polymer extension, are becoming a reality.

Synchrotron radiation methods are widely used to yield the highest possible resolution from low and high angle powder diffraction data within short timescales. The structural evolution of surfaces can be followed by x-ray surface diffraction and reflectivity.

As epitaxial growth occurs at a semiconductor surface it can be followed in real time to yield a precise determination of the atomic structure and density of the outermost layers of the crystalline solid. These weak effects rely principally on the unique combination of x-ray intensity and collimation—in other words, the brightness that synchrotron radiation sources provide. The collimation and control of the photon beam will also enable the crystallography of buried interfaces to be examined offering, for example, a direct understanding of the problems associated with doping, quantum confinement and the optical performances of solids. The use of XAFS enables the local atomic environment of a particular element in a solid or a liquid to be obtained independently of the remaining constituents. It does not rely on crystallinity and can be applied to the study of disordered materials including the growth of quantum dots and other nanometre sized particles. Our understanding of amorphous materials such as glass has been immensely extended by the application of XAFS methods which give the ability to observe the movement of selected cations, e.g., uranium, within the 'immobile' glassy environment.

Surface engineering is very important in materials research as it can correlate surface damage with the alteration to surface structures caused by a particular use. Probing the detailed chemistry of second and third row elements and the orientation of carbon, oxygen, nitrogen or molecular species at any surface demands high and controllable polarisation as well as intensity at soft x-ray energies.

Within the past few years significant advances have taken place in the exploitation of circularly polarised radiation for the study of magnetic materials using the technique of magnetic circular dichroism. The ability to identify separately the spin and orbital contributions of magnetic elements is enabling the formulation of new physics and opening the way to an understanding of how to create fundamental magnetic building blocks for sensors and storage devices and for detecting magnetic microstructures *in situ*.

An effective development has been the ability to apply several powerful synchrotron radiation methods simultaneously in real time, for example small and wide angle x-ray scattering, differential scanning, calorimetry, XAFS and x-ray diffraction to follow structural changes during the production of new materials.

Energy dispersive powder diffraction (possible only using white x-radiation from a storage ring) has already been used for kinetic studies on real systems, including the hydration and dehydration of cements and the crystallisation of waxes in fuel. Another example is the study of electrochemical processes which play such a vital role in everyday life in corrosion protection, paint technology, batteries, solid-state dielectrics and electrosynthesis. Synchrotron radiation provides selective access to the absorption edges of light elements (sulphur, aluminium, oxygen and others) which will enable studies, including electrolyte absorption, on dilute systems to be introduced and will permit experiments to be performed under real electrochemical conditions.

A great variety of surface chemical methods has evolved using the methods of near edge XAFS, surface XAFS, standing x-ray waves and photoelectron diffraction. Surface x-ray diffraction is at present the ideal way to study the structure of a near perfect crystalline surface and is excellent for the study of model chemisorbed systems. Polarised near edge XAFS studies offer a unique way to gain information on the symmetry and orientation of molecules adsorbed on surfaces. Many studies observing surface melting, freezing and crystallisation of long chain hydrocarbons on a wide range of structural surfaces have taken advantage of this technique.

Finally, in the materials science area, a fundamental understanding of the behaviour of catalysts at an atomic level is only possible using the techniques of x-ray absorption spectroscopy and x-ray diffraction. Although active sites in catalysts are usually dilute, XAFS can establish the structures, oxidation states and atomic environments of metals such as platinum, rhodium, molybdenum and nickel independently of crystallinity or homogeneity. Simultaneous measurements of short range order using XAFS and of long range order using x-ray diffraction are crucial to an understanding of the full surface chemistry of metals and substrates.

Molecular Science

In the study of the physics and chemistry of atomic and molecular systems, the ability to excite, ionise or multiply ionise molecules selectively, using single photon excitation, is extremely important.

Synchrotron radiation can be used for gas phase experiments to examine pathways for electronic decay and energy transfer in an enormous range of systems of atmospheric, photochemical and photophysical importance. Merged beam techniques (photons with ions) can be used; the resonant behaviour in vibrational modes of gases can be examined; high resolution photoelectron angle and energy distributions can be determined for almost all molecules. Fluorescence techniques can be used to study the decays of excited states of molecules, of their ions and of photo fragments.

Links with Industry

With the extension of the application of synchrotron radiation storage rings to the study of materials possibilities have evolved for industrial applications in areas such as chemical analysis, catalysis studies, pharmaceuticals development and new techniques in manufacturing. The initial exploitation by industry has been to employ synchrotron radiation sources for shallow lithography (semiconductors) and deep lithography (micromechanics using the LIGA process, figure 11). During the past ten years many compact synchrotron radiation sources have been developed for this rapidly evolving and potentially important area, as described in the following chapter.

The Future of Synchrotron Radiation

Around the world the number of dedicated storage ring sources for synchrotron radiation research and for commercial exploitation is continuing to grow. This growth is striking, especially because of the substantial capital costs associated with the design and construction of any new accelerator and of the even higher costs of the installation, equipment and operation of the many beamlines associated with each storage ring. A typical cost for the construction of a new source falls between £20M and £500M, with at least an equal amount being necessary for the construction of a reasonably complete portfolio of beamlines.

Recent years have seen many international collaborative activities in the field of synchrotron radiation research. The largest and best example is the European Synchrotron Radiation Facility (ESRF) in Grenoble, France, which involves a large number of

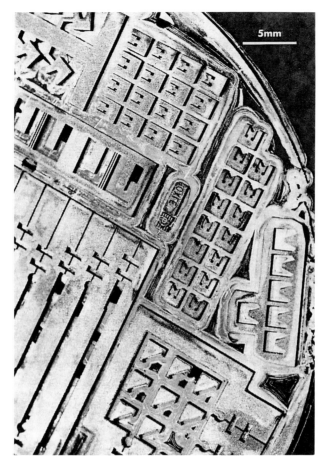

Figure 11. An example of micro engineering, i.e., a miniature engineering device produced with submicrometre surface finish by the process of deep x-ray lithography (LIGA), otherwise known as x-ray pattern printing. The device shown is a mould for passive optical switch components made using the SRS at Daresbury.

European nations including the UK. The number of facilities used for synchrotron radiation research, including those accelerators shared with high energy physics users, has grown from 3 in 1970 to 15 in 1980 and to 37 in 1990. In a recent catalogue of synchrotron radiation facilities it was predicted that by the end of the 1990s between 1000 and 1500 experimental beamlines and stations will be operating worldwide. This will represent a capital investment, in equipment alone, which will greatly exceed £1bn. Collectively, such sources and their associated experimental stations incorporate thousands of the highest quality, state of the art, optical elements including mirrors, gratings and crystals. This has already led to the creation of new industries to provide specialist ultrahigh vacuum compatible optics and precision instruments. With over 40 synchrotron radiation facilities already in operation and a further 20 facilities at the proposal or design phase, future developments are moving towards the production of super brilliance sources. Such devices, perhaps using FEL technologies, will incorporate all the tried and tested techniques developed during the past one or two decades to provide a beam size sufficiently small to obtain maximum information from submicrometre volumes of any material. Perhaps we now have in our grasp the ultimate tool to derive an understanding of the structure and function of every kind of biological and nonbiological material of significance to mankind.

Bibliography

Catlow C R A & Greaves G N (eds) "Applications of Synchrotron Radiation" Glasgow: Blackie and Son Ltd (1990)

Ebashi S *et al* (eds) "Handbook on Synchrotron Radiation, Volume 4" Amsterdam: North-Holland (1991)

Koch E-E (ed) "Handbook on Synchrotron Radiation, Volume 1" Amsterdam: North-Holland (1983)

Margaritondo G "Introduction to Synchrotron Radiation" New York: Oxford University Press (1988)

Marr G V (ed) "Handbook on Synchrotron Radiation, Volume 2" Amsterdam: North-Holland (1987)

Munro I H *et al* "World Compendium of Synchrotron Radiation Facilities" Orsey: The European Synchrotron Radiation Society (1991)

X-Ray Lithography

Alistair Smith
Oxford Instruments

T he inexorable demand for more and more computer memory has now continued unabated for several decades. Contrary to some predictions semiconductor memory chip producers have been required to introduce successive generations of DRAMs (Dynamic Random Access Memories), each with four times the capacity of the preceding generation, on a roughly three year cycle.

There are no signs of the process slowing down—in the consumer area, for example, the advent of products such as high definition television, or indeed any software requirement for 'solid-state video' where large numbers of pixels have to be addressed, will dictate large scale memory storage capacity. Software packages continue to grow and any changes to the user interface, involving speech and pattern recognition, could extend demand further.

The sequence is illustrated in figure 1, which charts the year of introduction for successive DRAM generations to date as well as a projection into the next century. Although personal computers with 16 Mbit RAM chips are now being advertised, this figure shows that manufacturers are just starting volume production of 64 Mbit devices. This probably reflects the fact that the bulk of memory is still directed to high end products and the initial output is used internally by the major producers.

While each successive device has four times the number of transistors, this is not achieved simply by increasing the physical size of the chip. The chips are indeed somewhat bigger, but the main factor is that the individual unit cell shrinks progressively. In other words, there are more and more elements produced in the same area of silicon. The so called 'critical dimensions' or 'linewidths'

X-RAYS: The First Hundred Years
Edited by Alan Michette and Sławka Pfauntsch © 1996 John Wiley & Sons Ltd

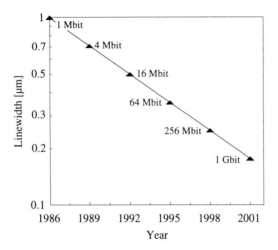

Figure 1. Year of introduction for DRAM generations.

corresponding to each generation are marked on the vertical axis of figure 1. For comparison, a human hair is around 50 μm in diameter while 0.1 μm corresponds to the dimensions of some viruses!

As well as allowing an increase in storage capacity, reducing the dimensions provides the additional important benefit of increasing the device speed—the shorter the 'gate length' of the transistor, the more rapidly it will switch. In fact, for certain types of device, such as microprocessors, this is the more crucial parameter. Intel's Pentium processor derives its increased performance through having smaller critical dimensions than those of a 486 chip, as well as by having more functional circuits.

The fabrication of devices is a complex process and involves many steps. We tend to imagine microchips as two-dimensional, but they are in fact three-dimensional in character with each layer, or level, requiring accurate placement relative to the preceding one. A typical sequence of events for producing just one level is illustrated in figure 2. A current generation memory device has around 15–20 levels, including all the metallisation layers that allow physical connection, and this involves hundreds of discrete process steps. Future generations will have still more levels.

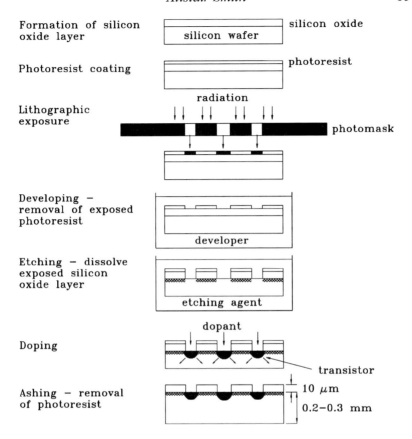

Figure 2. Typical process sequence in device fabrication.

The individual steps vary considerably in their criticality, but the most demanding are those which transfer the circuit patterns to the wafer, since, as explained above, these determine the device capacity and speed. This patterning step is called lithography and is viewed as the primary technology driver for the semiconductor device industry.

X-ray lithography is a development which may have significant impact in the future. However, in order to appreciate this some explanation of the lithography process and how it is currently performed optically, using visible light, is necessary. This is dealt with in the next section.

Optical Lithography

The lithographic process is analogous to photographic printing. In order to produce a photographic print, light is shone through a negative, and a lens is used to enlarge the image on to paper coated with a light sensitive emulsion. The paper is then chemically developed to produce the required positive.

In the case of lithography the circuit layout information, equivalent to the negative, which we wish to print and thereby transfer to the wafer is stored on a mask, and light is shone through this to project it on to the wafer which has been coated with a photosensitive material known as photoresist. Subsequent chemical development is similar in concept to the photographic analogue.

In order to achieve extremely high accuracy in lithography, the mask, a sort of stencil made of chromium lines deposited on a quartz substrate, is made larger than the final dimensions required, e.g., by a factor of 5, and the lens system operates in a demagnifying mode, in contrast to photography. Photographic enlargement and optical lithography are illustrated diagrammatically in figure 3.

To reduce costs a fabrication plant uses large silicon wafers—8" diameter is the current typical size, though there are plans to go to a diameter of 12". It follows that each wafer consists of a large number of devices, which can be processed together, and only at the end of the fabrication sequence are they sliced up to give individual chips. Thus, in the lithographic process each individual device has to be exposed sequentially and the wafer then moved, or stepped, to the next site.

The equipment which performs all of this is a so called reduction stepper. Again, there is an analogy with the store offering one hour processing of film, where the negatives are stepped through the enlarger one after the other. However, because of the layered nature of semiconductor devices, at each exposure site it is necessary to align the image relative to the preceding level using special reference marks. Typically, the accuracy of this alignment process should be about 10% of the critical dimensions of the device. Figure 1 shows that this implies an alignment precision measured in nanometres! Adding to that the fact that, from an economic standpoint, the throughput (or production rate) of wafers, each containing 50–100

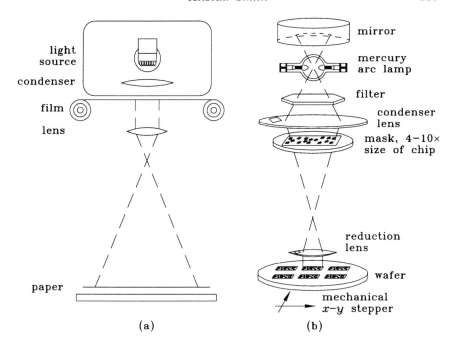

light
source

condenser

film

lens

paper

(a)

mirror

mercury
arc lamp

filter

condenser
lens

mask, 4–10×
size of chip

reduction
lens

wafer

mechanical
x–y stepper

(b)

Figure 3. Schematic diagram comparing (a) photographic printing and (b) optical lithographic printing.

devices, has to be in the range 30 to 60 per hour, it can be seen that these steps must be performed very rapidly. This is why current generation steppers cost > $5M!

The above is an essentially generic description of the process which uses light of visible wavelengths and which has been in use since the late 1970s. The sizes involved highlight a problem for the manufacture of devices with dimensions below 0.25 μm, as will be required early in the 21st century—the critical dimensions are at, or below, the wavelengths of visible, or even ultraviolet, light.

Until now the issue of reducing dimensions has been dealt with primarily by changing the wavelength of the light used as the source of illumination in the wafer stepper. Specific lines from a mercury discharge lamp, providing radiation at 436 nm (g-line) or 365 nm (i-line), are the workhorses in use for current production. Beyond these it will be necessary to switch to excimer lasers—krypton fluoride (at 248 nm) and argon fluoride (at 193 nm) are candidates.

Some clever optical tricks have also been developed which help to extend the viability of the conventional optical lithographic approach. Means have been developed for adjusting the illumination system to exclude all but zero and first order diffracted beams[1], and for modifying the mask such that it enhances the image in the photoresist, the so called phase shifting mask[2]. Both of these are themselves highly sophisticated topics, and are beyond the scope of this review.

The greatest difficulties associated with extending optical lithography further are essentially practical ones. Producing transmissive lenses at the wavelengths in question, and of the quality required to provide aberration and distortion free images over the field of view applicable to future chips, is immensely difficult and is as yet unproven at 193 nm. From an operational standpoint, however, an even greater problem is that of depth of field. This can again be best illustrated by comparing with photography where changing the lens aperture allows differential focusing—blurring the background in a portrait is a common technique. The problem in lithography is how

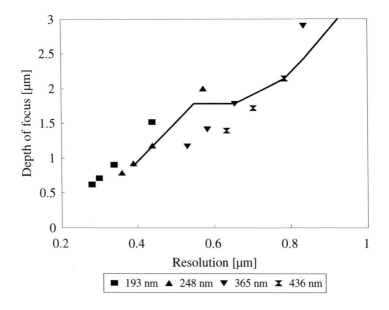

Figure 4. Depth of field versus resolution (linewidth) for g-line, i-line, KrF and ArF optical wavelengths. The line shows the progression in manufacturing to date.

to produce a lens with a large enough aperture to allow an adequate range of focus. Currently achievable depths of focus are quantified in figure 4. As explained earlier, chips are really three-dimensional in character, with topography of several hundreds of nanometres, i.e., the available depths of focus as shown in the figure are comparable with the physical variations in each layer on the chip. To make things worse, the resist coatings have poor reflective properties in the ultraviolet wavelength range. While this can be overcome by using multilayer coatings with a reflective outer layer, this simply adds further process steps and cost.

The above has described methods by which optical lithography is being extended to allow devices of smaller dimensions to be fabricated, i.e., at critical dimensions close to the wavelength of the light source. This evolutionary approach undoubtedly has appeal to those involved in manufacturing, but it is at the cost of ever increasing complexity and consequently has a potential financial impact in terms of yield.

The so called process window for optical methods is getting smaller and ultimately a more radical approach is required. X-ray lithography is seen as the most likely successor.

X-Ray Lithography

Conceptually, of course, the change to x-ray lithography is not really so radical—simply a shift to radiation of much shorter wavelengths. However, practically there are some significant alterations to both the equipment involved and the related processes.

Because of the large number of devices which have to be produced, a high flux of x-rays is imperative—arguably, the only practical solution is a synchrotron, the most powerful continuous source available.

We have seen in earlier chapters, and it is well documented in the literature[3], that synchrotron radiation is emitted by charged particles when they are accelerated and deflected by a magnetic field. The spectrum of radiation covers a broad range of wavelengths and, while invariant in shape, it has a critical energy dependent on the magnetic field and on the acceleration energy of the particles. The critical energy, or the corresponding critical wavelength λ_c, lies at the midpoint of the power spectrum. As will be seen this is important in determining the detailed design of an accelerator.

Clearly, the critical wavelength is one of the key parameters from which the design must begin. A complex set of related parameters determined by the lithographic process is involved. Diffraction effects, sensitivity of the photoresist and range of photoelectrons released in it are all factors to be considered, but the major determinant is probably the choice and form of the mask[4].

The differential absorption of x-rays, which makes them so attractive as a diagnostic tool both medically and industrially, means that masks of the type used in optical lithography are unsuitable for x-ray lithography. The use of masks with such a structure would result in an x-ray image with inadequate contrast. For x-rays a more complex mask, as illustrated schematically in figure 5, is needed. The substrate is a thin membrane of silicon, silicon carbide or silicon nitride, and the absorber, which blocks radiation, has to be of a high Z material such as gold, tungsten or tantalum[5]. In order to provide the required contrast, ideally around 10:1, the plated absorber layer has to be about 0.5 μm thick, while the membrane is typically 2 μm thick.

Glass absorbs x-rays and thus lens based optical systems are precluded. Focusing of x-rays using reflective mirrors is possible but quite difficult, and so a reduction scheme is not practical at the moment, particularly for higher energy x-rays. While the elimination of the large and expensive lenses needed for optical steppers is perhaps a benefit, it means that the mask itself must be used for 1:1 replication. Fabricating perfect masks economically at the lateral dimensions of the devices themselves, and with absorbers of the

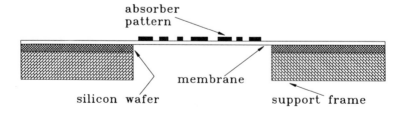

Figure 5. Schematic diagram of x-ray mask structure.

thickness required, is extremely difficult. Eliminating defects from these high aspect ratio structures is a major problem, and probably the single biggest issue to be resolved before the introduction of this technology can be effected.

The silicon membrane sets the upper spectral limit at photon energies of about 1800 eV because of absorption at the silicon K edge. The form of the beamline that transmits radiation to the stepper is described later, but an important issue is the interface which allows x-rays to emerge from the vacuum of the accelerator system to the wafer. As explained above, glass windows are not suitable and the exit windows are in fact made of thin pieces of beryllium, a material which is very absorbing for energies below 1000 eV. The synchrotron needs therefore to produce maximum x-ray flux in the 1000–1800 eV region, corresponding to wavelengths of 1.2–0.7 nm.

The preceding discussion suggests that a critical energy near to 1400 eV will maximise power within the lithography window. This conclusion was used in the design of the Helios machine built by Oxford Instruments for IBM. The spectrum and operating window of Helios are shown in figure 6.

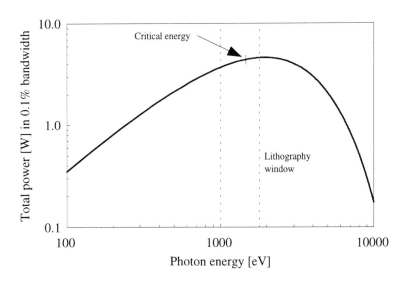

Figure 6. The power spectrum of Helios showing the preferred window for lithography and the critical photon energy.

The Helios Design

Having defined the spectral requirements the source itself must now be considered. Alan Wilson, the head of IBM's X-Ray Group in the 1980s, summarised the requirements for a synchrotron storage ring as follows[6]: "It should fit on a truck; plug into a wall socket; be reliable and available to operate 20 out of 21 shifts per week; have sufficient average output capable of sustaining a stepper throughput of >30 wafers per hour using an insensitive (~100 mJcm^{-2}) x-ray resist; be capable of being debugged/commissioned and assembled at the vendor, shipped intact to IBM, and rendered operational in a reasonable time at full specifications."

Storage rings based on conventional resistive magnets, such as the SRS at the Daresbury Laboratory, are physically large structures. Of course, how large depends on their energy and critical wavelength, but even a machine with spectral output centred on about 1500 eV would have a circumference of around 40 m. It would therefore fail to meet several of the criteria given above. Intact

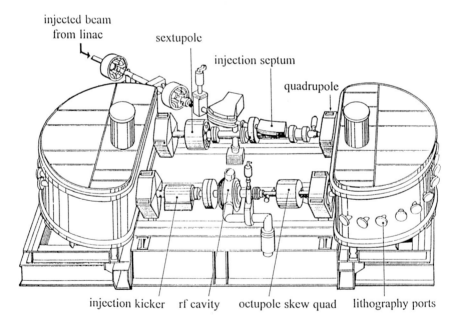

Figure 7. Schematic layout of the Helios storage ring. The dimensions of the main unit are roughly 6 m×2 m and the superconducting dipole magents are housed in the D-shaped chambers at either end of the ring.

transportation and speed of commissioning are real issues for a machine destined to be at the heart of a fabrication plant costing around a billion dollars.

The critical wavelength, in nanometres, is $\lambda_c = 20.74 R^{-2} B^{-3}$ where R is the bending radius, in metres, of the magnets and B is the magnetic field in tesla[7]. This shows that, for a fixed wavelength, the bending radius scales as $B^{-3/2}$ which implies that a considerable advantage may be gained by using superconducting magnets. In synchrotrons resistive magnets are limited to fields of about 1.3–1.4 T, whereas superconducting magnets can operate at much greater field strengths—the operating field of Helios is 4.5 T. Much smaller radii are thus possible, by a scaling factor of about 6, and it is feasible to shrink the machine to satisfy the mechanical conditions of the IBM specification.

Figure 7 shows the Helios system layout and figure 8 is a photograph of the unit following completion and prior to testing in Oxford.

In principle the simplest structure for a storage ring is to have just one circular magnet, and indeed a machine of this configuration

Figure 8. Photograph of Helios prior to testing in Oxford.

has been built in Japan[8]. However, all of the ancillary accelerator requirements still apply—the need for focusing and correction magnets and for a radio frequency (RF) cavity, and a means of injecting electrons. From a practical standpoint a two magnet race-track design, as adopted for Helios, is therefore preferable. The Helios ring comprises two 180° dipole magnets separated by two straight sections containing the ancillary equipment. When the synchrotron is operating at a field of 4.5 T, and the electrons are accelerated to 700 MeV, the x-ray spectrum of figure 6 is generated. This has a critical wavelength of 0.84 nm, corresponding to a critical energy of 1467 eV.

At the start of operation electrons are injected from a linear accelerator via the septum magnet in one of the straight sections. This pulsed magnet operates in combination with a second pulsed unit, the kicker magnet, on the other straight. These allow a significant current, 200–500 mA, to be built up quite rapidly at the injection energy of 100 MeV, and then the injector is switched off. The RF cavity is switched on and, with the magnet currents ramped synchronously, is used to accelerate the circulating electrons to the full 700 MeV energy. Of course, the cavity remains on to restore power lost through synchrotron radiation.

The power, in kilowatts, radiated during one orbit of the electron beam can be determined from the expression $P = 26.5 I E^3 B$ where I is the stored electron current in amps and E is the electron energy in GeV [7]. For Helios, at a stored current of 300 mA, the total power output is approximately 12 kW. The number of beamlines determines what fraction of the power is actually accessed. In the case of Helios the power to each of the 20 lines is 115 W. Compare that with a typical 8 kW hospital generator, where the power in this case relates to the input power—only about 2% of this results in useful x-rays and, because they are emitted isotropically, the power delivered to a patient is no more than a few watts.

The Helios design was the result of close collaboration between Oxford Instruments and the Daresbury Laboratory, the latter providing much of the accelerator physics expertise. The key engineering challenge was the design and fabrication of the superconducting magnets; a cross section of one of them can be seen in figure 9.

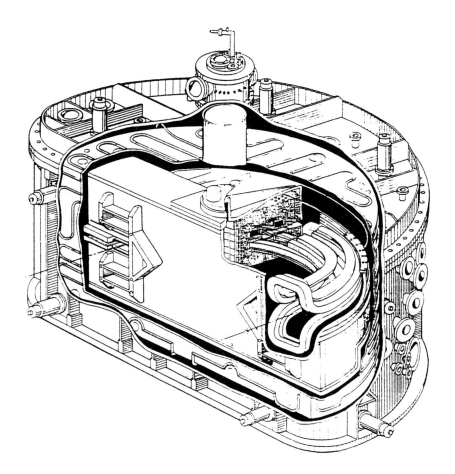

Figure 9. Cutaway view of dipole system cryostat showing coils within the 4.5 K structure, surrounded by an 85 K radiation shield and contained within a room temperature vacuum vessel.

Because of the completely open design, needed to allow the radiation to emerge, the superconducting coils are of a very complex shape, as is the support structure required to hold them accurately in place—at the full field current of 1.019 kA the closing force at the magnet aperture is equivalent to that generated by a mass of hundreds of tons.

In order to function, the superconducting coils must be cooled to about 4 K and therefore the whole inner structure is welded to form a cryostat containing the necessary liquid helium. A radiation shield, cooled via liquid nitrogen to 85 K, separates this from the external ultrahigh vacuum vessel which is, of course, at room temperature.

Under ideal conditions the stored beam would continue to circulate indefinitely. However, scattering by residual gas molecules in the vacuum chamber inevitably means that a progressive decay occurs—this is expressed in terms of a beam lifetime. Once the beam current drops too low it is necessary to dump the beam and repeat the injection sequence, causing the lithographic process to be held up. Long lifetimes are therefore highly desirable and dictate a requirement for extremely high vacuum conditions, which were achieved by coupling the vacuum enclosure to the liquid helium cryostat, effectively forming the walls of the vessel into a massive cryopump, a so called cold bore configuration. As a consequence the vacuum quality is superb and lifetimes longer than 24 hours are routinely achieved.

The synchrotron radiation is emitted in a continuous fan around each magnet arc. This is shown schematically in figure 10 which indicates the horizontal angular width $\Delta\theta$ accepted by each individual beamline; for Helios $\Delta\theta=60$ mrad. One of the characteristics

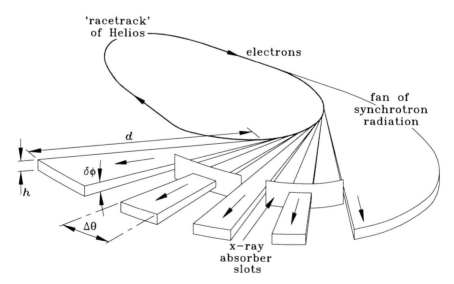

Figure 10. Emission of synchrotron radiation around a magnet arc from the racetrack electron orbit of Helios.

of synchrotrons is that the vertical angle $\delta\phi$ is very small[9]. For Helios $\delta\phi\approx1$ mrad, which results in a beam height of 10 mm for a 10 m long beamline.

To obtain high throughput, each individual exposed area should be as large as possible. The horizontal beam dimension satisfies this requirement, and by scanning the beam vertically, using mirrors, a field size of about 25 mm×25 mm is achieved. A field of this size is suitable for exposing two or three memory devices at a time. The generalised arrangement of a lithography station is illustrated schematically in figure 11. The focus range of such a beamline is about 40 μm; an attractive feature of x-ray lithography is its ability to print images over extreme topography.

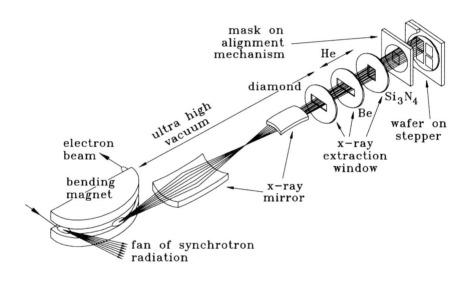

Figure 11. Schematic of synchrotron radiation lithography beamline.

One of the obvious differences between x-ray and optical configurations is that the x-ray wafer plane is vertical, while in optical lithography the wafers are always horizontal, so that the steppers have to operate in different orientations. Whereas optical steppers each have their own light source, a ring such as Helios can simultaneously support up to 20 x-ray steppers. The cost of the source is thus spread over a large number of systems and cost of ownership models[10] indicate that, in capital terms, the two are approximately equivalent. Of course in the event of a system failure to the storage ring all units would be halted, which explains the need in the IBM criteria for excellent reliability. A rupture of just one of the beryllium windows, for example, could threaten the whole system and must be avoided. Thus, a lot of the equipment in the beamline (not shown in figure 11) is associated with protecting the vacuum integrity of the storage ring.

Helios Performance

The first Helios system was designed and constructed during the period 1988 to 1991 and tested at Oxford Instruments. Following achievement of the key specifications it was then transported intact to IBM's Advanced Lithography Facility in East Fishkill, New York. Figure 12 shows the machine, in the cradle used to transport it,

Figure 12. Photograph of Helios arriving at IBM's East Fishkill facility.

arriving at its destination. Beams were stored within 8 weeks of arrival, and the ring was performing to its full specifications just a couple of months later.

In fact the machine surpassed all its required specifications comfortably, as summarised in table 1. At the output power indicated, exposure time per 25 mm×25 mm field is about 2–4 s, consistent with the specified throughput requirements.

Equally important are the operational statistics that define the system reliability. In the parlance of semiconductor tools, the key measure is the availability, or uptime. Figure 13 shows this for the period 1992 to mid 1995—the average uptime is greater than 98%.

Table 1. A comparison of the design characteristics with the actual performance of Helios.

Parameter	Design value	Achieved to date
electron energy	700 MeV	700 MeV
dipole field	4.5 T	4.5 T
injection energy	200 MeV	200 MeV & 100 MeV
injection current	10 mA	20 mA
injection pulse length	100 ns	100 ns
electron beam current		
at injection energy	200 mA	650 mA
at full energy	200 mA	297 mA
bending radius	0.519 m	0.519 m
total x-ray power	8.2 kW	12 kW
electron beam size		
radial	$\approx$1.4 mm	1.4 mm
vertical	$\approx$1.4 mm	1.4 mm
vertical divergence	$\approx$0.6 mrad	
radio frequency	499.7 MHz	499.7 MHz
base pressure	5×10^{-10} torr	3×10^{-10} torr
operating pressure	3×10^{-9} torr	5×10^{-9} torr
beam lifetime	>5 hours	26 hours at 200 mA
yearly uptime	>90%	>98%

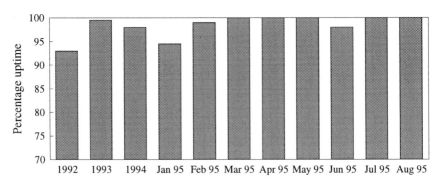

Figure 13. Helios uptime record.

This is the first demonstration of the use of a synchrotron in industry—all previous systems have been located in research environments. While synchrotrons are indeed complex and sophisticated assemblies, it is clear that with suitable design it is perfectly feasible to envisage storage ring x-ray sources as industrial tools for the 21st century.

References

(1) B J Lin "Off Axis Illumination—Working Principles and Comparison with Alternating Phase Shifting Masks" *Proc. SPIE* **1927** 89 (1993)

(2) M D Levenson "Wavefront Engineering for Photolithography" *Physics Today* **46**(7) 28 (1993)

(3) E-E Koch *et al* "Synchrotron Radiation—a Powerful Tool in Science" *Handbook on Synchrotron Radiation* vol 1a (E-E Koch, ed) Amsterdam: North-Holland p 1 (1983)

(4) S D Hector *et al* "Simultaneous Optimization of Spectrum, Spatial Coherence, Gap, Feature Bias and Absorber Thickness in Synchrotron Based X-Ray Lithography" *J. Vac. Sci. Tech.* **B11** 2981 (1993)

(5) J P Silverman *et al* "Performance of a Wide Field Flux Delivery System for Synchrotron X-Ray Lithography" *J. Vac. Sci. Tech.* **B11** 2976 (1993)

(6) A D Wilson "X-Ray Lithography in IBM, 1980–1992, the Development Years" *IBM J. Res. Develop.* **37** 299 (1993)

(7) S Krinsky *et al* "Characteristics of Synchrotron Radiation and of its Sources" *Handbook on Synchrotron Radiation* vol 1a (E-E Koch, ed) Amsterdam: North-Holland p 65 (1983)

(8) H Yamada "Commissioning of Aurora—the Smallest Synchrotron Light Source" *J. Vac. Sci. Tech.* **B8** 1628 (1990)

(9) M J Poole and M N Wilson, to be published in "Proximity X-Ray Lithography" (W G Waldo, ed) New Jersey: Noyes Publications (1996)

(10) S Ishihara, paper presented at XEL Conference, Osaka, Japan, July 1995

X-Ray Astronomy

Ken Pounds
Particle Physics & Astronomy Research Council and Leicester University

S ome 50 years after x-rays were discovered by Wilhelm Conrad Röntgen, the Sun was established to be a strong x-ray source. Fifty years on, more than 60 000 extraterrestrial sources of x-rays, most being many times more powerful than the Sun, have been detected by instruments carried above the Earth's opaque atmosphere. Their increasingly detailed study has led to a new and powerful means of observing the Universe, x-ray astronomy.

The Sun as an X-Ray Source

Fifty years ago the true nature of the Sun's corona was a mystery. Observable then only on the rare and brief occasions when the Moon lay in direct line of sight, thereby eclipsing the millionfold brighter photosphere, the corona shines mainly in sunlight scattered towards the Earth. In addition the coronal light included a number of emission lines whose atomic origins had defied identification, leading to the proposal of a new element, 'coronium' , unknown on Earth. In fact a paper, largely missed by a world at war, had been published in the early 1940s by the Swedish spectroscopist Edlén[1] who showed the most prominent of these coronal emission lines to be due to the common element iron, albeit in stages of ionisation up to Fe XIV. Neutral iron, i.e., the form in which the element exists under normal circumstances, is labelled Fe I by spectroscopists; Fe II is singly ionised iron, i.e., an atom of iron with one electron removed and Fe XIV is iron with 13 of its electrons removed. Immediately postwar, other pieces of evidence fell into place, so that the large radial extent of the corona, the absence of photospheric absorption lines in the scattered light, and the high degree of ionisation, were all

X-RAYS: The First Hundred Years
Edited by Alan Michette and Sławka Pfauntsch © 1996 John Wiley & Sons Ltd

explained in terms of a remarkably high coronal temperature, of the order of a million degrees Celsius. Soon the British astrophysicist Fred Hoyle added the proposal that such a hot plasma would be a copious source of x-radiation, thereby explaining the further mystery of the origin of the Earth's ionosphere[2] (x-rays incident on the upper atmosphere causing ionisation).

Direct confirmation of Hoyle's prediction, and the beginning of a powerful new means of observing the solar corona, came in 1948 with the flight of a photographic x-ray sensor, on a modified V-2 rocket, by scientists from the US Naval Research Laboratory (NRL). Exposure to the Sun, above the absorbing layers of the Earth's atmosphere, resulted in blackening of film shielded from the visible sunlight by metal foil[3]. Throughout the 1950s the NRL group, led by Herbert Freidman, carried out a series of rocket flights, with film and Geiger counter detectors, confirming the intensity of the Sun's x-radiation and its strong variability, associated with solar flares[4,5]. In parallel, a more thorough theoretical basis was established in a series of papers by Elwert[6–9].

My own involvement in this research began in 1959 when, as a research student at University College London, I had the opportunity to send both film and proportional counter detectors to 150–200 km above the Woomera Desert in Australia in the new British sounding rocket, Skylark[10]. Over the next decade Skylark, equipped with an accurate Sun pointing attitude control system, provided teams at UCL and Leicester University (where I moved in 1960) with an excellent platform to study the x-ray emission of the Sun.

During this period, the first x-ray images[11] and resolved x-ray spectra[12] were obtained and the era of satellite borne solar studies began[13]. This work has advanced to the point where, today, the Japanese Yohkoh satellite, a collaboration with US and UK astronomers, is producing a continuous stream of high quality coronal images (figure 1). The ability to now observe the corona continuously and to determine its changing structure, temperatures, mass motion and chemical abundance, is a direct consequence of the high degree of discrimination against the more luminous solar photosphere which, as a consequence of its much lower temperature, emits essentially no x-rays.

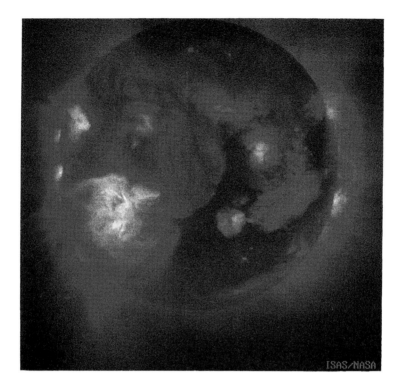

Figure 1. An x-ray image of the solar corona taken with a grazing incidence telescope/CCD camera combination on the Yohkoh satellite.

Cosmic X-Ray Astronomy—the Early Days

The opportunity brought by the emerging space era for astronomers to observe extraterrestrial sources of x-rays naturally engaged the thoughts of theorists and experimenters during the late 1950s. However, prospects for detecting 'cosmic' sources beyond the solar system did not seem good, given the great distances to even the nearest stars and the still more distant, and rare, supernova remnants that were predictable candidates as emitters of x-rays. Nevertheless, experiments were actively discussed[14]. In the UK a collaboration between UCL and Leicester put forward a proposal in 1961 to fly an array of modest sized x-ray telescopes on the third in a new series of NASA spacecraft, the Orbiting Astronomical Observatories. Other, more ambitious, plans were being developed by Freidman and colleagues in the USA.

All this was overtaken by a dramatically successful rocket experiment carried out in June 1962 by a group of physicists from the Massachusetts Institute of Technology and a private company, American Science and Engineering. Originally inspired by the great Italian physicist Bruno Rossi, then at MIT, a powerful collaboration of theorists and experimenters had emerged determined to make a serious attempt at opening up the field of 'cosmic x-ray astronomy'. In the event, the 1962 flight of the Aerobee 150 rocket from White Sands, New Mexico, had the more modest aim of detecting scattered or fluorescent x-rays from the sunlit Moon. However, any possibility of detecting lunar x-rays was swamped by a remarkably strong source in the general direction of the galactic centre[15]. At an x-ray flux of 5 photons $cm^{-2} s^{-1}$ this signal, if from a nearby star, would represent an x-ray luminosity $10^7 - 10^8$ times that of the Sun.

Subsequent rocket borne experiments, by the AS&E group in October 1962 and June 1963 and by the NRL group in 1963, not only confirmed the initial detection as from a point source, at that time unidentified, in the constellation Scorpius, but also detected other cosmic x-ray sources in Cygnus and in Taurus. The latter was identified with the Crab Nebula supernova remnant by the NRL group in an ingenious July 1964 rocket borne observation timed to coincide with a lunar occultation of the nebula[16].

Other discoveries followed thick and fast. The AS&E/MIT group flew a new payload in March 1966 in which the conventional slatted detector collimator was replaced with grids of wires forming the so called 'modulation collimator'. This device, invented at MIT by Minoru Oda, later to become a leader of the emerging Japanese x-ray astronomy programme, enabled the position of the intense source Scorpius X-1 to be greatly refined[17]. This led quickly to its identification with an unusual blue star[18], of visible magnitude 13, implying the existence of stellar objects with x-ray emission 1000 times that in the visible spectrum! In the same year, 1966, the first extragalactic x-ray source was detected, from the powerful radio galaxy M87[19].

Back in the UK these developments were being followed with great interest and we had noted the opportunity afforded by Skylark and access to the Woomera rocket range to extend the US led surveys into the southern hemisphere. In 1967 the Leicester group began a series of Skylark flights which, over the following 10 years, carried out the first full surveys of the southern sky[20], observed the first x-ray nova (in Centaurus), successfully flew a crystal spectrometer/polarimeter to observe Scorpius X-1[21] and obtained the (still) most precise x-ray source position, locating the source GX 3+1 to 0.3 arc second accuracy using the lunar occultation technique[22].

From 1970, however, the field was once more dominated by Riccardo Giacconi and his colleagues in the AS&E group, with the successful launch from Kenya of the first dedicated x-ray astronomy satellite, Uhuru, which means 'freedom' in Swahili, a name chosen in recognition of the launch on Independence Day in Kenya.

The Satellite Era

The launch of Uhuru, on December 12, 1970, took x-ray astronomy a major step forward. The simple but powerful payload, consisting of an array of proportional gas counters of effective area $700\,cm^{-2}$, scanned a great circle of the sky coincident with the spin plane of the spacecraft[23]. The spin axis could be accurately pointed in any direction giving an all sky observing capability. The combination of an ultimate point source sensitivity of the order of 10^{-4} of the flux of Scorpius X-1, of star and Sun sensors allowing x-ray sources to be located to a fraction of a degree, and of signal timing to 0.1 s led directly to quantitative and qualitative advances. Over the next couple of years the all sky survey increased the number of known x-ray sources to 161[24]. More significantly, new and unexpected classes of source were discovered, in particular the extremely powerful 'x-ray binaries' (of which Scorpius X-1 was found to be one example) and the extended emission from previously unsuspected hot gas filling the vast space within clusters of galaxies.

Figure 2 reproduces the map of the x-ray sky derived from the first complete (3U) Uhuru Catalogue. Although occupying most parts of the sky, a concentration of the brightest sources along the galactic plane indicates their origin within the local (Milky Way) galaxy. Among this group are Scorpius X-1, the Crab Nebula, Cygnus X-1 and Centaurus X-3. The latter two are both x-ray binary systems, in which the intense x-ray flux is produced by matter falling

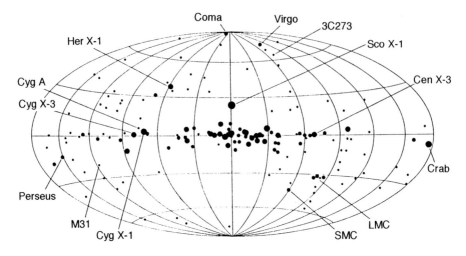

Figure 2. A map of the sky in galactic coordinates from the Uhuru (3U) Catalogue. The location of 161 cosmic sources is shown, with the size of each dot proportional to the logarithm of the x-ray flux.

from a nearby companion star onto a black hole or neutron star. In the case of Centaurus X-3 the detection of regular 4.8 s pulses confirmed the compact object to be a neutron star, spinning with that period. By determining the arrival times of individual pulses, the Uhuru team were also able to measure the second order shift in this period due to Doppler effects of the 2.087 day binary star period[25]. The precise fit of the arrival times to the expected Doppler sine curve is shown in figure 3. Additional information on the inclination of the Centaurus X-3 binary system, provided by the regular two day eclipse of the x-ray source, allowed a very precise determination of the system's orbital velocity, radius and mass function (which depends on the masses of the two components and the angle at which the orbital plane is viewed). Cygnus X-1, in contrast, showed no regular x-ray pulsations, its x-ray flux varying on all timescales down to the 0.1 s resolution limit of Uhuru. The implication that, in this case, the compact accreting object was a 'featureless' black hole was supported by the identification of the stellar companion as a massive 9th magnitude BO supergiant[26], yielding a mass of at least five times that of the Sun ($M_\odot$) for the compact object (significantly above the theoretical mass limit of $3M_\odot$ for a neutron star).

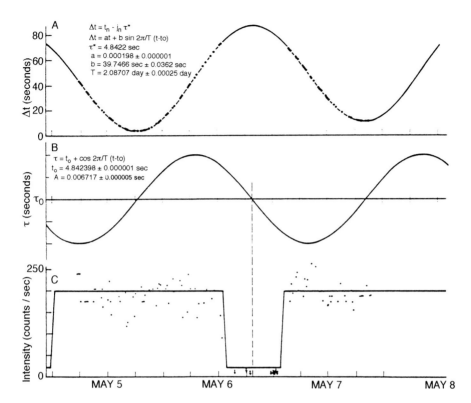

Figure 3. Uhuru observations of the eclipsing x-ray binary Centaurus X-3 in May 1971. Panel A plots the time difference in arrival of 4.8 s pulses from that for a constant period. Panel B illustrates the dependence of the observed pulse period on the phase binary cycle. Panel C shows the mean intensity during successive 100 s spin passes, clearly demonstrating the eclipsing of the x-ray source.

Uhuru's ability to carry out extended observations of individual sources, and in particular of the variability that was proving to be a feature of many cosmic x-ray sources, led to the majority of the bright galactic objects being established in the x-ray binary class. In total, it was found that the luminous x-ray binary systems accounted for some 90% of the total x-ray emission of our Galaxy. In contrast, supernova remnants, representing the second most frequent type of luminous galactic x-ray source (and the only class to be predicted in advance), were found to be weaker and less numerous in the Uhuru survey, only relatively young ($<2\times10^4$ years) supernovæ emitting detectable flux in the Uhuru (2–10 keV) energy band.

Prior to the launch of Uhuru, only three extragalactic x-ray sources had been confidently identified, these being associated with the giant nearby radio galaxies M87 and Centaurus A and the powerful quasar 3C273. The Uhuru (3U) survey contained some 60 sources at high galactic latitude. These included several members of the Local Group of galaxies, including the Large and Small Magellanic Clouds and M31, the Andromeda Nebula. Several active galaxies including two quasars were also found. However, the largest subset of extragalactic sources was, again, of a totally unexpected type, namely clusters of galaxies. In the optical band, many galaxies are found to cluster in apparently gravitationally bound groups of tens to hundreds. The remarkable finding of Uhuru was that several of the more pronounced clusters exhibited powerful x-ray emission, extending across the whole span (often many megaparsecs) of the cluster. My own association with the initial realisation that the extended x-radiation we were seeing from the Virgo cluster was associated with the cluster itself, rather than being from a combination of its brighter constituent galaxies, represents my outstanding memory of several months on leave at AS&E in the early days of the Uhuru mission[27].

Following the lead of Uhuru, several other dedicated satellite missions were put into orbit during the 1970s. Next in line, and one of the most productive of these later missions, was the British Ariel V satellite launched, like Uhuru, into an equatorial orbit from Kenya. Ariel V carried several instruments including a sky survey array from the Leicester group, viewing in the spacecraft spin plane, and a proportional counter detector, viewing along the spin axis, from UCL. These two instruments were the most productive over the orbital lifetime of Ariel V, which ran from launch in 1974 until satellite re-entry in 1980. The sky survey sensitivity was quite similar to that of Uhuru, although the rather better source positions led to a significant number of new identifications. The most notable advance in this respect was in the area of the extragalactic sources where I had the personal satisfaction of being able to rule out the UHGLSs, a mysterious class raised by the initial Uhuru survey[28]. The new Ariel V data showed that these previously 'unidentified high galactic latitude sources' were yet more clusters of galaxies, together with a new class of powerful x-ray emitter associated with Seyfert galaxies. For me, that discovery led to a long association with the study of Seyfert galaxies, in which x-ray observations have revolutionised our understanding of the active nucleus and its immediate environment in these powerful extragalactic objects.

Looking back, the most dramatic observations from Ariel V were the discoveries of several remarkable transient sources such as another class of extragalactic x-ray emitter, the BL Lacertids[29]. However, the most remarkable transient was AO620-00, discovered in August 1975 during the first European Conference in Astronomy held in Leicester. This object remains today the brightest cosmic x-ray source ever seen. Adding to the drama was the then unique facility by which data from Ariel V could be transmitted to a ground station on Ascension Island and then displayed in Leicester within 24 hours. As the conference proceeded, AO620-00 grew in brightness, to exceed that of the Crab Nebula, then of Scorpius X-1 itself. Subsequently, over several weeks, the x-ray flux faded again[30], alleviating concern in some parts of the media that the world was under threat. Optical study of the now faint (19th magnitude) star identified with AO620-00 has since shown it to be a strong candidate for an x-ray binary system containing a black hole[31]. Historical optical records, furthermore, suggest that this system undergoes transient mass transfer at intervals of 58 years. Future astronomers will surely be on the lookout for a repeat event in 2033!

The Modern Era

The third era in the development of x-ray astronomy, and the period in which it became a major observational discipline of astronomy, comparable in importance to the optical and radio bands, began in 1979 with the launch of the Einstein Observatory[32]. Remarkably, this was a project again led by Giacconi's group, albeit now moved to the Center for Astrophysics at Harvard. The unique feature of the Einstein Observatory was the first orbiting of a large imaging x-ray telescope dedicated to the study of cosmic sources. The telescope resolution, of the order of 5 arc seconds, brought the triple advantages of direct imaging of extended sources, enhanced sensitivity arising from the signal to noise gain of the focused beam, and the precise location (and hence relative ease of identification) of the many new sources to be seen.

In the event, Einstein proved every bit as productive as expected, identifying several thousand new sources, resolving individual sources in nearby galaxies, yielding structural information on extended emission from many supernova remnants and galaxy clusters, and detecting active galaxies to high redshift[33]. In addition, the detection of many normal stars, thereby connecting with the early theoretical conjectures of the 1950s, showed x-ray emission to be a powerful means of studying stellar activity as a function of age, rotation rate and type[34].

After three years of successful operation the Einstein mission ended, and then began a long period in which US x-ray astronomers, until then very much the dominant community, have taken a back seat to researchers in Europe and Japan. From 1983–6 the European Space Agency's EXOSAT mission held the stage with a combination of large proportional gas counters and an array of focusing telescopes[35]. A feature of EXOSAT was its orbit, the 200 000 km apogee allowing up to three days viewing of individual sources uninterrupted by Earth shadowing. This operational feature led to some of the most important results from EXOSAT, including unique light curves of eclipsing binary systems[36] and of a number of Seyfert galaxies[37]. The latter were found to be variable on timescales down to tens of minutes (figure 4), a truly remarkable result for the emission from a galaxy clearly showing the x-rays to arise in a tiny region (a massive black hole?) in the heart of the galactic nucleus. A major impact of EXOSAT's success was the boost to European x-ray astronomy and British astronomers, from a position of relative strength, gained the greatest share of observing time.

In 1987 the centre of activity moved to Japan, with the launch of the GINGA satellite. The main instrument on board GINGA was the Large Area Counter (LAC), built jointly by research groups in Nagoya, Tokyo and Leicester, offering the largest detector array (about 4000 cm^2) ever placed in orbit[38]. This large and sensitive array allowed further advances in the study of variability and in broadband spectroscopy of the few hundred brightest cosmic sources over the 2–20 keV band[39]. The data archive resulting from many

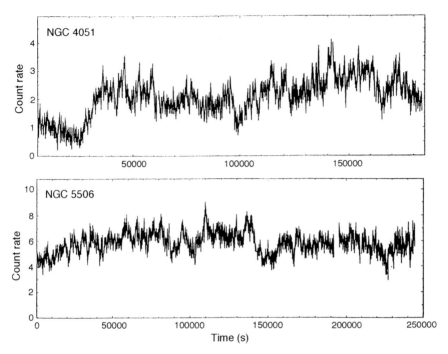

Figure 4. Extended EXOSAT observations of two Seyfert galaxies showing continuous, rapid and large amplitude variability.

extended observations over the orbital life of GINGA (until its re-entry in November 1991) is still in regular use by astronomers worldwide. A particular advance in my own area of interest, Seyfert galaxies, arising directly from the additional spectral sensitivity of the LAC, was the detection of reprocessing effects in the observed spectrum, indicating the presence of dense clouds or streams of gas (probably associated with the primary accretion process) in the close vicinity of the postulated black hole[40].

The 1990s have so far seen the launch of two more x-ray satellites, the German ROSAT (an abbreviation of Röntgensatellit) in June 1990 and the Japanese ASCA in February 1994. Both missions represent further major advances in this now mature field of astrophysical research. ROSAT carries a German x-ray telescope, similar to Einstein but with improved resolution and sensitivity[41], and a smaller telescope built in the UK and optimised for the first complete survey in the extreme ultraviolet (XUV) waveband[42]. A feature of the ROSAT mission was a systematic all sky survey with both telescopes yielding, respectively, over 60 000 x-ray sources (down to 10^{-8} of the flux of Scorpius X-1) and almost 500 sources of XUV radiation[43]. From early 1991 the ROSAT mission has continued, until now, operating in a pointed phase analogous to the observational mode of a ground based telescope. To date several thousand detailed studies of a wide range of cosmic x-ray sources have been carried out for astronomers around the world.

The ASCA project has brought another qualitative advance, this time in the diagnostic study of individual sources[44]. Its imaging telescopes feed detectors of higher energy resolution than the gas proportional counters used hitherto, a pair of photon counting charge coupled device (CCD) detectors providing spectra of up to tenfold better spectral resolution than available before. The significance of the ASCA data is heralding a new era in x-ray astronomy, in which spectroscopy will allow diagnostic studies of the high temperature/ high energy universe comparable to the well proven techniques available in the optical and infrared bands.

The Scientific Impact of X-Ray Astronomy

It is a truism in astronomy that any substantial enhancement in observing capability, sensitivity, resolution or waveband, brings major new discoveries. The experience of 33 years of x-ray observations have fully justified that belief, as expressed by Bruno Rossi back in 1960, when he explained his own motivation for encouraging colleagues at MIT and AS&E as being "a deep seated faith in the boundless resourcefulness of nature, which so often leaves the most daring imagination of man far behind".

The ROSAT sky survey detected significant x-ray emission from the hot component of the interstellar medium, most types of normal star, and all classes of galaxy, in addition to compact binaries, galaxy clusters and active galaxies at all redshifts. In other words, pretty well the whole of astronomy!

However, the x-ray data is far from 'more of the same', and frequently adds a new dimension to the physical understanding of an object or system, primarily because of a different regime of temperature, particle energy or field strength. Many examples could be chosen, from the new insights on the internal magnetic fields of normal stars via the energy transported into their coronæ, to the propagation of shocks in supernova remnants from the temperature and density distribution of residual hot gas, and the gravitational mass and cooling flows in galaxy clusters from similar x-ray image and spectral data. For brevity, I will choose two other areas, physically related, for more detailed examples of the major impact the new x-ray data are having — our understanding of x-ray binaries and of active galactic nuclei.

The most luminous x-ray binary systems, which as noted earlier dominate the x-ray emission of the Galaxy, involve mass transfer from a 'normal' star onto a neutron star or black hole. The x-ray luminosity is often close to that where the accretion rate is limited by radiation pressure (the 'Eddington limit'), or $10^{31}M$ W, where M is the mass of the compact component in units of the solar mass. In 'low mass' binaries, i.e., where the companion star is of the order of the Sun's mass, this x-ray flux often accounts for over 99% of the total luminosity of the binary star system. X-ray data, combined with optical information on the companion star, can be used to determine the radius and mass of each star, limited only by uncertainty on the system inclination (unless it is an eclipsing system). X-ray data provide a direct measure of the rotation rate of a neutron star companion, while detection of cyclotron line emission yields a measure of the magnetic field strength (often as high as 10^8 T)[45]. In a few cases, thermonuclear explosions occurring from a buildup of hydrogen on a neutron star surface have been used to derive a measure of the stellar radius (typical radii for neutron star masses of $1.4M_\odot$ are about 7 km), on the assumption of an optically thick (blackbody) emitter[46].

The detection of a gravitational redshift in an emission or absorption feature arising close to the neutron star surface is another way of determining the mass to radius ratio. Given the mass and radius, models of the equation of state of the neutron star itself can be constrained[47]. In the study of black hole systems the major thrust to date has been in obtaining unambiguous lower limits of the mass of the compact companion star. Since all current physical models of neutron stars have mass limits of $<3M_\odot$[48], such evidence is generally accepted to provide the strongest proof of the existence of a black hole.

The physical similarity of the latter class of x-ray binary with active galactic nuclei is that these extreme luminosity objects are also widely believed to be energised by accretion onto a black hole[49], albeit in this case of mass in the range 10^6–$10^9 M_\odot$. The extended EXOSAT observations (see figure 4) show clearly the remarkably fast and large amplitude variability of the x-ray emission from Seyfert galaxies. On light travel time arguments this demonstrates directly the small scale of the emission region, comparable to that of the inner solar system. A scaling of the Eddington limited luminosity to a $10^6 M_\odot$ black hole yields an x-ray power of 10^{37} W, similar to that found for many Seyfert galaxies; however, other evidence suggests active galactic nuclei may typically be less well fed, with a typical accretion rate more like 0.01–0.1 times the Eddington limit.

The essential reason why x-ray observations are of primary value in studying active galactic nuclei is that the accretion process

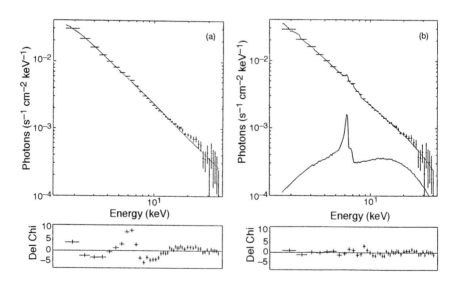

Figure 5. (a) A power law fit to a Seyfert spectrum observed by GINGA, together with the residuals between the data and the fit. (b) Reduction of the residuals by addition of a 'reflection' component.

results directly in the main gravitational energy release, which is inversely proportional to the radius, occurring very close to the black hole. Here the temperature of the accreting gas rises to tens of millions degrees Celsius (equivalent to kiloelectron volts in energy terms), where x-rays provide the main cooling mechanism (and emission). Thus, x-ray timing and spectral data are the most direct probes of the 'central engine' in active galactic nuclei[50].

The diagnostic potential of x-ray spectral data from recent missions such as GINGA and ASCA is well illustrated by the case of Seyfert galaxies. Figure 5 shows a Seyfert spectrum, obtained using GINGA, in which features superimposed on the intrinsic power law reveal the presence of dense matter, probably associated with accretion, via fluorescence and electron scattering effects[40]. Figure 6 illustrates how the improved spectral resolution of the CCD detectors on ASCA has allowed the iron K fluorescent line to be resolved, showing a broad redshifted wing apparently arising in the strong gravitational field close to the central hole[51].

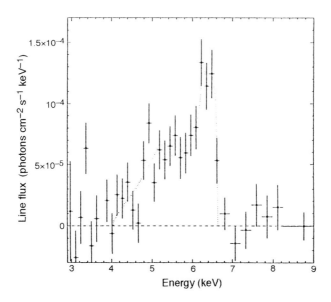

Figure 6. The Fe K fluorescent line profile from the Seyfert galaxy MCG-6-30-15 observed by the CCD camera on ASCA. The line is extremely broad with a pronounced 'red' wing attributed to Doppler and gravitational shifts close to a central black hole.

The Future

By the end of the present century three major new x-ray satellite projects should be in orbit—the Russian Spectrum-X in 1997, NASA's AXAF mission in 1998 and ESA's XMM spacecraft in 1999. The powerful imaging telescopes should finally realise the full potential of using the Röntgenstrahlen from cosmic sources to probe the physical nature of our Universe.

Acknowledgements

To the extent that this account of the development of x-ray astronomy reflects the author's personal experiences over the last 35 years, the crucial assistance of many colleagues in the UK and abroad is gratefully acknowledged.

References

(1) B Edlén "An Attempt to Identify the Emission Lines in the Spectrum of the Solar Corona" *Ark. Mat. Astr. Fys.* **28B** 1 (1941)

(2) F Hoyle & D R Bates "The Production of E-Layer" *Terr. Magn. Atmos Elect.* **53** 51 (1948)

(3) E Durand *et al* "Analysis of the First Rocket Ultraviolet Solar Spectra" *Astrophys. J.* **109** 1 (1949)

(4) H Friedman "The Solar Spectrum Below 2000 Ångstroms" *Ann. Geophys.* **11** 174 (1955)

(5) T A Chubb *et al* "Measurements Made of High Energy X-Rays Accompanying Three Class 2^+ Solar Flares" *J. Geophys. Res.* **65** 1831 (1960)

(6) G Elwert "The Continuous Emission Spectrum of the Solar Corona in the Far UV and the Adjacent X-Radiation" *Z. Naturforsch.* **7a** 202 (1952)

(7) G Elwert "Ionisation and Recombination Processes in a Plasma, and the Ionisation Formula for the Solar Corona" *Z. Naturforsch.* **7a** 432 (1952)

(8) G Elwert "The Soft X-Radiation from the Undisturbed Solar Corona" *Z. Naturforsch.* **9a** 637 (1954)

(9) G Elwert "The X-Ray Radiation of the Solar Corona and Hot Coronal Condensations" *Solar Eclipses and the Ionosphere* (W J G Beynon & G M Brown, eds) London: The Pergamon Press p 167 (1956)

(10) K A Pounds & P J Bowen "A Simple Rocket Borne X-Radiation Monitor—its Scope and Results of an Early Flight" *Monthly Not. Roy. Astron. Soc.* **123** 4 (1962)

(11) P C Russell & K A Pounds "Improved Resolution X-Ray Photographs of the Sun" *Nature* **209** 490 (1966)

(12) K Evans & K A Pounds "The X-Ray Emission Spectrum of a Solar Active Region" *Astrophys. J.* **152** 319 (1968)

(13) K A Pounds "The Solar X-Radiation Below 25 Å" *Ann. Geophys.* **26** 555 (1970)

(14) A I Berman (ed) *Proc. Conf. X-Ray Astronomy* Boston: Smithsonian Astrophysical Observatory (1960)

(15) R Giacconi *et al* "Evidence for X-Rays from Sources Outside the Solar System" *Phys. Rev. Lett.* **9** 439 (1962)

(16) S Bowyer *et al* "Lunar Occultation of X-Ray Emission from the Crab Nebula" *Science* **146** 912 (1964)

(17) H Gursky *et al* "Measurement of the Location of the X-Ray Source Sco X-1" *Astrophys. J.* **146** 310 (1966)

(18) A R Sandage *et al* "On the Optical Identification of Sco X-1" *Astrophys. J.* **146** 316 (1966)

(19) E T Byram *et al* "Cosmic X-Ray Sources, Galactic and Extragalactic" *Science* **152** 66 (1966)

(20) B A Cooke & K A Pounds "Further High Sensitivity X-Ray Sky Survey from the Southern Hemisphere" *Nat. Phys. Sci.* **229** 144 (1971)

(21) R E Griffiths *et al* "Fluctuations in the X-Ray Intensity of Sco X-1" *Nat. Phys. Sci.* **229** 175 (1971)

(22) A F Janes *et al* "Identification of GX3+1 from Lunar Occultations" *Nature* **235** 152 (1972)

(23) R Giacconi *et al* "An X-Ray Scan of the Galactic Plane from Uhuru" *Astrophys. J.* **165** 127 (1971)

(24) R Giacconi *et al* "The Third Uhuru Catalogue of X-Ray Sources" *Astrophys. J. Supp.* **27** 37 (1974)

(25) E Schreier *et al* "Evidence for the Binary Nature of Centaurus X-3 from Uhuru X-Ray Observations" *Astrophys. J. Lett. Ed.* **172** 69 (1972)

(26) B L Webster & P Murdin "Cygnus X-1—a Spectroscopic Binary with a Heavy Companion?" *Nature* **235** 37 (1972)

(27) E Kellogg *et al* "The Extended X-Ray Source at M87" *Astrophys. J. Lett. Ed.* **174** 65 (1972)

(28) K A Pounds "An X-Ray Map of Deep Space" *Ann. New York Acad. Sci.* **302** 361 (1977)

(29) M J Ricketts *et al* "X-Ray Transient Source at High Galactic Latitude and Suggested Extragalactic Identification" *Nature* **259** 546 (1976)

(30) M Elvis *et al* "Discovery of Powerful Transient X-Ray Source A0620-00 with Ariel V Sky Survey Experiment" *Nature* **257** 656 (1975)

(31) J E McClintock & R A Remillard "The Black Hole Binary A0620-00" *Astrophys. J.* **308** 110 (1986)

(32) R Giacconi *et al* "The Einstein/HEAO 2/X-Ray Observatory" *Astrophys. J.* **230** 540 (1979)

(33) Y Avni & H Tananbaum "On the Cosmological Evolution of the X-Ray Emission from Quasars" *Astrophys. J. Lett. Ed.* **262** 17 (1982)

(34) R Pallavicini *et al* "Relations Among Stellar X-Ray Emission Observed from Einstein, Stellar Rotation and Bolometric Luminosity" *Astrophys. J.* **248** 279 (1981)

(35) N E White & A Peacock "The EXOSAT Observatory" *Mem. Soc. Astron. Ital.* **59** 7 (1988)

(36) A N Parmar *et al* "The Discovery of 3.8 Hour Periodic Intensity Dips and Eclipses from the Transient Low Mass X-Ray Binary EXO 0748-676" *Astrophys. J.* **308** 199 (1986)

(37) K A Pounds & I M McHardy in *Physics of Neutron Stars and Black Holes* (Y Tanaka, ed) Univ. Acad. Press. p 285 (1988)

(38) M J L Turner *et al* "The Large Area Counter on GINGA" *Publ. Astron. Soc. Japan* **41** 345 (1989)

(39) Special issue of *Publ. Astron. Soc. Japan* **41** (1989)

(40) K A Pounds *et al* "X-Ray Reflection from Cold Matter in the Nuclei of Active Galaxies" *Nature* **344** 132 (1990)

(41) M R Sims *et al* "XUV Wide Field Camera for ROSAT" *Opt. Eng.* **29** 649 (1990)

(42) J Trümper *et al* "X-Ray Survey of the Large Magellanic Cloud by ROSAT" *Nature* **349** 579 (1991)

(43) J P Pye *et al* "The ROSAT Wide Field Camera All Sky Survey of Extreme Ultraviolet Sources. II. The 2RE Source Catalogue" *Monthly Not. Roy. Astron. Soc.* **274** 1165 (1995)

(44) Y Tanaka *et al* "The X-Ray Astronomy Satellite ASCA" *Publ. Astron. Soc. Japan* **46** 37 (1994)

(45) K Makishima & T Mihara in *Frontiers of X-Ray Astronomy* (Y Tanaka and K Koyama, eds) Univ. Acad. Press p 23 (1992)

(46) W H G Lewin *et al* "X-Ray Bursts" *Space Sci. Rev.* **62** 223 (1992)

(47) J Van Paradijs *et al* "A Very Energetic X-Ray Burst from 4U 2129+11 in M15" *Publ. Astron. Soc. Japan* **42** 633 (1990)

(48) G Baym & C Pethick "Physics of Neutron Stars" *Ann. Rev. Astron. Astrophys.* **17** 415 (1979)

(49) M J Rees "Black Hole Models for Active Galactic Nuclei" *Ann. Rev. Astron. Astrophys.* **22** 471 (1984)

(50) R F Mushotzky *et al* "X-Ray Spectra and Time Variability of Active Galactic Nuclei" *Ann. Rev. Astron. Astrophys.* **31** 717 (1993)

(51) Y Tanaka *et al* "Gravitationally Redshifted Emission Implying an Accretion Disk and Massive Black Hole in the Active Galaxy MCG-6-30-15" *Nature* **375** 659 (1995)

The Development and Applications of Laser Produced Plasma X-Ray Sources

Ciaran Lewis
Queen's University Belfast

L aser produced plasmas have been available for laboratory study since shortly after the demonstration of the first optical lasers about 35 years ago. The early development of Q-switched megawatt lasers provided coherent optical radiation which could be focused to power densities of about 10^{10} W cm^{-2}. This is enough to convert an irradiated target material into a plasma, i.e., to remove electrons from the atoms of the material, thereby ionising it. After 30 years of international research into many aspects of laser plasmas, plans are now well developed in the USA to build the National Ignition Facility, a laser capable of providing 500 TW of optical power (2 MJ in 4 ns) in the early years of the 21st century. Parallel developments in short pulse laser technology will lead to petawatt lasers (1 kJ in 1 ps) in the near future. The target interaction processes depend on the characteristics of the driving laser—wavelength, duration, irradiance, polarisation, etc. However, in general, laser plasmas are regions of high temperature and high energy density and hence can function as bright sources of electromagnetic radiation with emission well into the x-ray part of the spectrum.

Incoherent X-Rays

When a high power laser beam is focused onto a target there is rapid ionisation of surface atoms. This enables most of the laser pulse energy to be absorbed by the freed electrons near to charged ions by a process known as inverse bremsstrahlung. The absorbed energy is distributed into the target material, mainly via thermal conduction through electron collisions but also by radiation from the plasma. The heated material is driven (ablated) away from the target surface in a rapid expansion caused by high pressures in the ablation zone. Typical plasma parameters can be gauged from a laser delivering 100 J of energy in a 1 ns pulse at a wavelength of about 500 nm, focused to a 1 mm diameter spot on an aluminium target. The incident irradiance is approximately 10^{13} W cm^{-2} causing ablation about 1 μm deep into the target. The electrons and ions of the heated plasma reach a temperature of nearly 500 eV (5×10^6 K) and in the densest part, near the ablation front, the electron density exceeds 4×10^{21} cm^{-3}, equivalent to about 0.5% solid density. The pressure here is very high, about 10 Mbar (10^6 J cm^{-3}) and it is this plasma region which radiates most of the x-rays.

Conversion efficiencies from input laser energy to emitted x-rays in an energy band of, say, 0.5–5 keV vary with exact conditions and target atomic number but can be as high as several tens of percent. Highest efficiencies are observed using short wavelength (ultraviolet) lasers. The emitted x-rays are said to be thermal as their energies represent the plasma temperature and they reflect conditions of ionisation and recombination. The x-rays are incoherent, are emitted into 2π sr, and have a duration similar to the heating laser pulse. Their spectrum generally consists of discreet lines from transitions between bound ionic states superimposed on regions of continuum due to free-free (bremsstrahlung) and free-bound transitions. Only in high atomic number high density plasmas does the spectrum resemble a Planckian source with blackbody emission characteristics reflecting the plasma radiation temperature.

Spectral Features

In a hot plasma, where ionisation and recombination are competing, it is usual to have several ionisation stages coexisting. However, only two or three stages tend to have significant fractional abundances, which dominate the spectral output from the plasma. As the temperature of a plasma increases the ionisation stages present change in a complex manner, since to be stripped from the parent atom consecutive electrons require different amounts of energy. For example, it is easy to ionise lithium- or sodium-like ions as the valence electrons are loosely bound. Adjacent helium- and neon-like ions are much harder to ionise due to their tightly bound closed shell structures. Thus, in plasmas, due to lack of sensitivity to temperature changes closed shell ions with 2 (helium-like), 10 (neon-like) and 28 (nickel-like) bound electrons tend to dominate the ionic composition of the plasma and hence the spectral output. Excitation and further ionisation within these closed shells leads to excited states radiatively decaying to final states within the shells. The resulting spectra are referred to as K, L and M shell spectra respectively.

The simplest K shell spectra belong to hydrogenic ions whose basic spectrum is that of the hydrogen atom with the wavelengths reduced by a factor Z^2, where Z is the number of protons in the nucleus. Examples are shown in figures 1(a) and (b) for hydrogenic carbon (C VI) and aluminium (Al XIII). The former shows clearly the Balmer series (principal quantum number transitions $n \rightarrow 2$, with $n > 2$) and the latter shows the Lyman series (principal quantum number transitions $n \rightarrow 1$). Both spectra also show emission lines from helium-like ions, although the principal series member ($n=2 \rightarrow n=1$) for Al XII at 7.76 Å has not been recorded. The weaker lines to the long wavelength side of many of the main (resonance) lines are known as satellite lines. They arise since some ions have an additional excited electron which slightly modifies the transition energy of the decaying electron. These satellites can be useful as diagnostics for the plasma density and temperature.

The aluminium spectrum also shows a short wavelength continuum, arising from free electrons recombining into bound states. The continuum slope can be used to determine the electron temperature. Finally, the resonance series are seen only up to transitions from about $n=6$. This is due to broadening of the transition lines from perturbations of the high lying ionic levels through electric microfields from plasma electrons and ions. The highest series member which can be clearly resolved gives a measure of the density of the perturbers and similar information is available from the shape of individual well resolved lines. Thus, even the simplest K shell spectra contain much subtle detail. Furthermore, the spectra are

space and time dependent so that measurements with spatial and
temporal resolution at the micrometre and picosecond level are often
required, for example to validate computational modelling of the
complex interaction between laser beam and plasma.

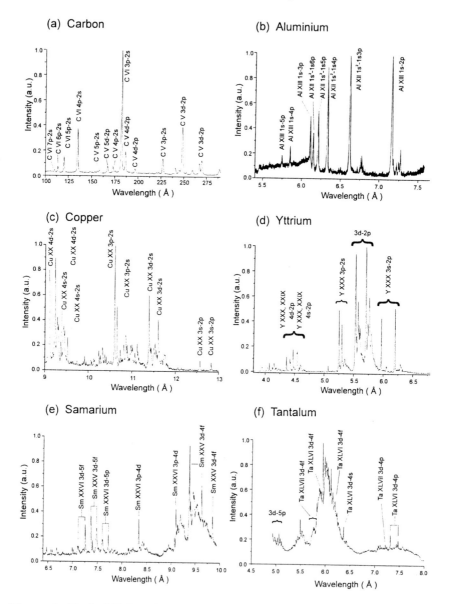

Figure 1. Typical K, L and M shell spectra from laser produced plasmas heated
with a variety of lasers.

L shell spectra are more complicated as shown in figures 1(c) and (d) for copper and yttrium targets. The spectra are still dominated by two ion stages, fluorine- and neon-like, but the satellite structures (from doubly excited neon- and sodium-like ions) linked to the main resonance transitions are more complex and usually not well resolved. The denser line spectra arise from the splitting of the energy levels in each principal quantum number shell and from the presence of ions with excited inner shell (e.g., 2s) electrons. For L shell spectra $\Delta n=0$ transitions (i.e., no change in principal quantum number) may also occur in addition to the $\Delta n \geq 1$ ones. This is significant for x-ray laser development, as will be seen later.

M shell spectra, as seen in figures 1(e) and (f), are even more complex due to the very large number of initial and final states available for a given $\Delta n=1$ transition. For high Z targets relativistic effects contribute to the splitting. Usually each transition gives a band of densely packed lines known as an unresolved transition array. Although the spectra tend to be dominated by nickel-like ions, it is common to have several ions with higher and lower charge coexisting and the corresponding unresolved transition arrays from these are shifted in energy but barely resolved. In plasma conditions (which can be controlled with the drive laser) where satellite lines are abundant and intense the net effect is to fill in the gaps and to produce a broad band $(\Delta\lambda/\lambda \approx 0.1)$ quasi continuous spectrum.

The range of photon energies which can be excited from the K, L, M and higher shells is illustrated in figure 2 for $\Delta n>0$ transitions.

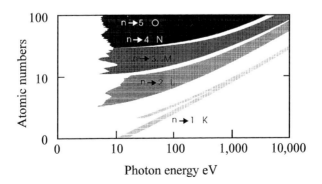

Figure 2. The photon energy ranges possible in emission spectra from laser plasmas if there is sufficient ionisation of the initial target element.

Table 1. Laser parameters used for generating the spectra of figure 1.

Spectrum	Wavelength [μm]	Energy [J]	Pulse length [ns]	Focal spot size [mm]	Irradiance [W cm^{-2}]
Carbon	0.53	0.5	5	0.5	$\approx 5 \times 10^{10}$
Aluminium	0.53	10	0.05	0.04	$\approx 10^{16}$
Copper	0.53	0.5	5	0.05	$\approx 5 \times 10^{12}$
Yttrium	1.06	600	1	0.1×20	$\approx 3 \times 10^{13}$
Samarium	1.06	600	1	0.1×20	$\approx 3 \times 10^{13}$
Tantalum	0.27	5	0.06	0.03×0.3	$\approx 10^{15}$

The ability to populate the plasma with a given ion state depends on the temperature achieved with the laser heating pulse. To generate the brightest output at a specified photon energy usually requires some optimisation of target material and focusing conditions. As a rule of thumb, for target irradiance around 5×10^{13} W cm^{-2} the spectral brightnesses of L, M and N shell emitters peak for elements with atomic numbers near 30, 60 and 90 respectively. These ions have ionisation potentials of about 2 keV.

The wide range of parameters available for generating laser plasma x-ray sources can also be gauged from the actual conditions used to produce the spectra of figure 1, as shown in table 1.

Applications

Laser produced plasma x-ray sources are very versatile because the conditions of the interaction between laser and target can be tuned to provide a wide range of source characteristics, allowing a diversity of applications. Most applications are related to basic science but some are showing commercial possibilities due to the relative compactness of a laser based system compared to alternative sources.

Probably the most general application of laser plasmas has been as laboratory sources for spectroscopy of highly ionised atoms. The L shell spectrum of the impurity molybdenum is seen in Tokamak plasmas which, through magnetic confinement, seek to emulate the fusion processes occurring in stellar interiors. Much of the emission spectroscopy associated with such impurity lines has been unravelled with the aid of laser plasmas. In addition, many studies have been related to astrophysical problems in identifying the spectral signatures of stars. For example, the L shell spectra of iron and nickel are seen in the solar corona and in solar flares. Recently, absorption spectroscopy has contributed to our understanding of radiation transport and opacity effects in stellar atmospheres. The term 'laboratory astrophysics' is often used to describe the application of laser plasmas to this type of basic research.

Equations describing the state of dense matter can be tested by various experimental techniques. The dynamics of crystal lattice planes, as they respond to the rapid compression associated with strong shock waves caused by a focused laser beam, have been measured using time resolved diffraction of x-rays from an independent but synchronised laser plasma. The technique known as EXAFS (extended x-ray absorption fine structure) has been used in studies of ion position correlations in amorphous materials shocked to high density, low temperature states. This also relies on synchronised laser pulses to shock the material and to provide the x-ray probe. EXAFS monitors small oscillations, caused by nearby atoms, in the energy dependent K shell absorption of x-rays.

Spatially extended laser plasma x-ray sources, filtered with broadband metal absorbers, have been widely used for diagnostic radiography of various other laser plasma related structures. This has been common in studies of inertial confinement fusion, where a symmetric arrangement of high power lasers is used to generate a high ablation pressure on the surface of a hollow shell target containing deuterium and tritium gases. The sustained pressure causes the submillimetre diameter shell to implode, heating and compressing the encapsulated gas to thermonuclear conditions. Synchronised, pulsed radiography, using pinholes, zone plates or reflective optics, gives information on the symmetry of the imploding structures at selected times. Typically, 100 ps pulses of x-rays with energies of several keV are used. Alternatively, time resolved images, using long pulses (typically a few nanoseconds) for backlighting and either x-ray streak cameras or fast framing cameras (with <100 ps frames) for detection, can be used to monitor the dynamics of these very transient events. Self emission of x-rays from these targets has also been widely used to measure density and temperature. This is often achieved by introducing tracer elements to provide optically thin emission which depends on the environment.

The broadband absorption techniques can be extended in a general way using the setup illustrated in figure 3. Here, advantage is taken of a very small (<5 μm) short duration (typically <50 ps) laser plasma source of x-rays to cast a projection shadowgraph of an object, which is another laser plasma in figure 3. Simultaneously, the source and absorbed spectra are both displayed by means of a dispersion element located in front of the film or CCD detector. It may be a Bragg crystal for keV x-rays or a grazing incidence spectrometer for lower energies. This powerful technique allows essentially monochromatic opacity measurements ($\Delta\lambda/\lambda \approx 10^{-4}$ with a crystal) to be made with 5 μm spatial resolution in two dimensions and in 50 ps frame times which can be scanned shot by shot through a much longer event. An example is shown in figure 4 where the probed object is a static, solid 100 μm diameter sphere which has been radiographed with 3 keV x-rays from a bismuth plasma back-

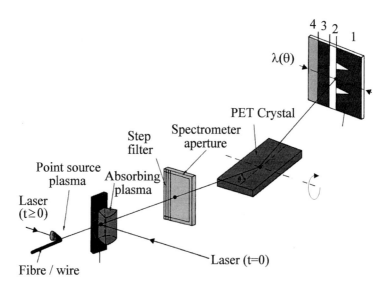

Figure 3. Arrangement for point projection spectroscopy. In this case the sample is the coronal plasma of a laser plasma heated in a line focus.

━ 100μm ━

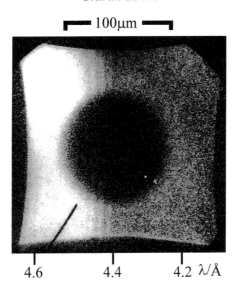

4.6   4.4   4.2 λ/Å

Figure 4. Point projection radiograph of a 100 μm diameter tantalum sphere at about 3 keV showing <10 μm resolution. The backlighter was a bismuth coated, 25 μm diameter gold wire and the discontinuity in the M shell spectrum is due to the K edge of a chlorine containing filter.

lighter. In this example the absorption was saturated; however, quantitative measurements can be obtained from imploding shells and the data can be used to validate hydrodynamic codes simulating the event. In plasma physics, the word 'code' is used to describe a computer program used to carry out complex simulations.

An example of this technique in the study of the corona of a rapidly expanding plasma is seen in figure 5. Here, an aluminium foil is irradiated by a 100 ps laser pulse and the expansion is recorded at about 50 ps time intervals for about 1 ns using absorption of a samarium backlighter. The frame shown indicates the position of hydrogenic and helium-like ions 150 ps into the expansion. The superimposed broad line spectrum is from self emission on the same resonance series transitions from the much larger aluminium plasma. Absorbance measurements allow absolute ion densities to be estimated and this provides a good test of modelling for this type of interaction, which is appropriate to recombination x-ray laser schemes.

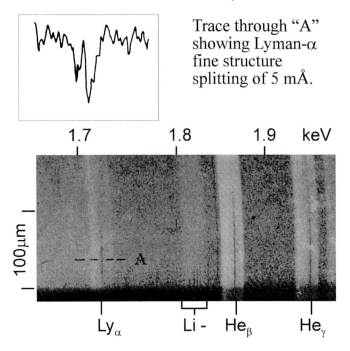

Trace through "A" showing Lyman-α fine structure splitting of 5 mÅ.

Figure 5. Point projection radiograph of an aluminium coronal plasma 150 ps after the peak of the short laser pulse which heated the target. Absorptions from the ground state of hydrogenic and helium-like ions are present as well as inner shell absorptions (lithium-like). The plume front is expanding at about 5×10^7 cm s^{-1}.

A final example, where the probed object is not a directly heated laser plasma, is illustrated in figure 6. The target, a thin layer of aluminium sandwiched between two plastic layers, is heated by the x-rays from an adjacent gold plasma. This indirect heating forms the aluminium plasma at much higher density than possible with optical radiation and hence the plasma should be close to local thermodynamic equilibrium. The tamped expansion of the aluminium plasma leads to fairly homogeneous conditions with well defined density and temperature. The absorption spectrum can be determined by extremely complex atomic physics codes; such a simulation is compared to experiment in figure 7. This shows the absorption spectrum of backlighter photons which promote K shell electrons into open L shells of several coexisting ionisation stages of aluminium. The space dependent spectrum—the x-ray flux heating the

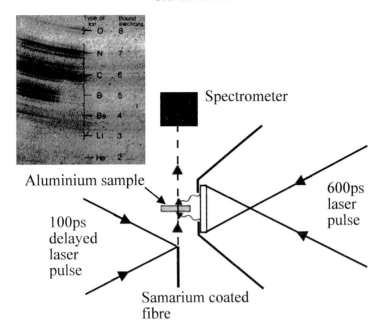

Figure 6. Arrangement for point projection spectroscopy of a radiatively heated sample. The inset shows 1s–2p absorption lines in the open L shell of aluminium ions.

sample is weaker further away from the gold plasma—is seen as an inset to figure 6. The code can be used to estimate the spatially varying temperature which determines the fractional abundance of the ion stages and hence the absorption spectrum signature. Such techniques can be applied to, for example, to the calculation of opacity in stellar atmospheres.

Not all applications of laser plasmas are spectroscopic. The emission from yttrium or tungsten plasmas, filtered to provide about 2–4 nm radiation in nanosecond pulses, is used for single shot contact microscopy of hydrated biological samples, such as algae and muscle cells, contained in wet sample holders in the target vacuum chamber. The range of photon energies used, known as the 'water window', is special in that it corresponds to a region of high contrast in absorption between carbon dominated structures (proteins, etc.) and oxygen dominated structures (water). Soft x-ray

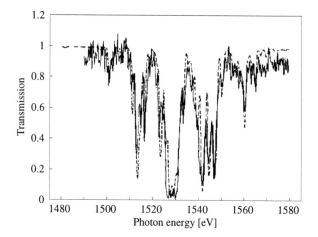

Figure 7. A detailed atomic physics calculation to simulate the measured absorption spectrum of a radiatively heated aluminium sample. The code assumes homogeneous conditions at a temperature of 40 eV and a density of 13 mgcm^{-3}.

fluxes of 80 mJcm^{-2} are needed to expose PMMA photoresist in contact with the samples and, after development, the resist is read with an atomic force microscope. Spatial resolutions of better than 50 nm have been demonstrated.

Recent experiments with very high incident intensity beams ($>10^{18}$ Wcm^{-2} in picosecond pulses), have shown that very high energy electrons and photons can be generated in a plasma which is heated at near solid density due to lack of expansion time during the pulse. Table-top, terawatt lasers operating at 10 Hz with 150 fs pulses tightly focused onto solid targets such as tantalum have shown that it is possible to produce high fluxes of very hard x-rays (up to 100 keV) in picosecond pulses. These can be used for high magnification projection radiography of massive objects with extremely high spatial resolution.

Laser plasmas can also be used as sources for x-ray lithography. At the UK Central Laser Facility trains of eight 7 ps pulses from a KrF amplifier ($\lambda=0.25$ μm) deliver 140 mJ of energy and are used to heat a copper target which emits L shell radiation at about 1.2 keV. The system runs at 50 Hz and the radiation has been used to expose photoresist on a wafer after passing through a mask consisting of narrow gold stripes supported on a thin silicon nitride membrane. With the wafer-mask combination at 125 mm from the plasma, a 5 minute exposure is enough to reproduce structures with dimensions of 0.18 μm. Such structures have been incorporated into silicon FET devices to define working gate electrodes (i.e., good I-V characteristics) as shown in figure 8 and may have important consequences in future 1 Gbit DRAM technology.

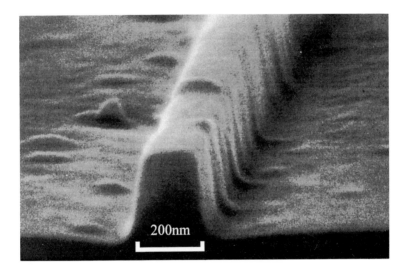

Figure 8. A 200 nm silicon gate electrode structure made using laser plasma based x-ray lithography.

Incoherent x-rays can also replace the optical lasers previously used to implode the shell targets employed in inertial confinement fusion research. This is achieved by focusing optical laser beams onto the inside walls of a nearly closed metal canister, typically gold, with millimetre dimensions (for kJ/ns lasers), with a centrally mounted target (figure 9). The device is known as a 'hohlraum'. The emission, absorption and re-emission of x-rays from the plasmas formed lead to an almost uniform radiation bath within the cavity. This high degree of uniformity is crucial for efficient compression, as it avoids the 'hot spots' intrinsic to coherent optical drivers, resulting from the limited number of beams and interference

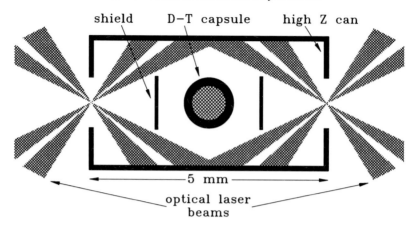

Figure 9. Arrangement for indirect drive of a deuterium-tritium capsule within a hohlraum. Cones of optical laser beams are azimuthally staggered to maximise isotropy.

between them. The process is referred to as 'indirect drive' and a hohlraum with a temperature of 200 eV (2×10^6 K), radiating at about 10^{14} W cm^{-2}, can generate an ablation pressure of about 100 Mbar on the shell surface. Direct drive using visible wavelength lasers delivering 10^{15} W cm^{-2} can develop similar ablation pressures, but indirect radiation drive scales as the fourth power of temperature. It offers the possibility of thermonuclear fusion ignition on the National Ignition Facility with 2 MJ energy into 1 cm hohlraums.

Coherent X-Rays

As indicated earlier, laser produced plasmas emit radiation characteristic of their state of equilibrium. This is determined by the balance between ionisation and recombination, which may not be the same in different regions of the plasma and is usually time dependent. At sufficiently high densities the plasma can be close to local thermodynamic equilibrium, when all collisional and radiative processes come into equilibrium with their inverse processes. In this limit the plasma emission will be that of a blackbody radiator. At low densities, such as found in the plume of hot plasma expanding away from the ablation zone, the balance is between processes of collisional ionisation and excitation and those of radiative recombination and spontaneous radiative decay. This leads to coronal equilibrium, which is similar to the situation found in the solar corona, and emission rates from the plasma are largely determined by the collisional excitation rates.

Usually, the laser produced plasma is in an intermediate state known as collisional radiative equilibrium. In this, high energy bound states (which are energetically close to their neighbouring states) tend to be in collisional equilibrium with each other while the populations of lower lying states are dominated by radiative decay. The high bound states that are statistically populated in accordance with the plasma temperature and those lower states that are not are separated by the so called collision limit. It is between energy levels that straddle this divide that one can seek to establish a population inversion in the plasma ions. This happens when a system is pushed so far out of equilibrium that upper states become more populated than lower states, the reverse of the normal case. Achieving this condition is a basic requirement to produce amplification of radiation at a frequency matched to the transition energy between the inverted levels; the process is called 'pumping'.

A population inversion between two excited levels sets the scene for amplification. A resonant photon interacting with an ion whose excited electron is in the upper state can create a new and identical photon by stimulating the electron to decay to the lower level, while a photon interacting with an ion whose electron is already in the lower state can be absorbed. For a net gain in photon numbers a population inversion is therefore needed. The degree to which the population densities can be pumped out of equilibrium determines the gain coefficient α of the active medium. For a plasma of length L the incident photon flux is amplified with a gain of $G \approx \exp(\alpha L)$. Typically, $L \approx 2$ cm and $\alpha \approx 3$ cm^{-1} and so $G \approx 400$.

If the energy gap between the inverted levels is large enough, as found in highly charged ions in a laser plasma, the amplifying photons can be in the x-ray part of the spectrum. Soft x-ray laser action of this type has now been demonstrated in laser produced plasmas at many different wavelengths. The shortest reported wavelength of 3.5 nm was obtained using a gold target whose atoms were stripped of 51 electrons, resulting in nickel-like ions. The brightest reported emission is at a wavelength of 15.5 nm from an yttrium target whose atoms were stripped of 29 electrons giving neon-like ions. In the latter case the emission has a high degree of coherence (i.e., the photons emitted are related to each other in an organised manner characteristic of a laser beam) and the beam brightness corresponds to that of a blackbody emitter with a temperature of several GeV (10^{13} K!), which is about a million times larger than the actual plasma temperature. Thus laser plasmas are emitters of both thermal and nonthermal radiation in the soft x-ray region. The following paragraphs outline how it is possible to move the plasma from one mode to the other.

Since the early 1960s many workers have published papers discussing the feasibility of extending the performance of lasers to progressively shorter wavelengths. It was clear that this presented a major technological difficulty because of the problem of pump power scaling. The radiative lifetime of the upper state of an inverted transition rapidly shortens as the potential lasing photon energy ($h\nu$) increases. Hence the pump power needed to sustain an inversion big enough to provide a given gain coefficient scales very adversely with energy—as $\nu^{9/2}$ under typical conditions. Thus terawatt, or larger, pump powers were likely requirements for soft x-ray lasers. Nevertheless, several pumping schemes were developed with the incentive that, if the lasers could be pumped successfully with a constant efficiency, their output brightnesses would also scale as $\nu^{9/2}$. The main schemes associated with the use of laser plasmas are photo-pumping, recombination pumping and collisional pumping.

In photopumping, a bright source of photons, often the line emission from an auxiliary laser plasma, is used to resonantly excite, through absorption, an upper level which is thus preferentially pumped relative to a lower level. A transient inversion can be set up between the levels before the system reaches a new equilibrium with the radiation source. These schemes are analogous to those visible lasers which are pumped by flashlamps or other lasers, e.g., dye lasers or diode lasers. Unfortunately, photopumping relies on wavelength coincidences between pump lines and pumped transitions and requires pump powers corresponding to plasma radiation temperatures on the limit of what is currently possible. No convincing demonstration of amplification based on these schemes has been reported to date but photopumping is often considered as an aid to recombination or collisional pumping.

In recombination pumping, a plasma is prepared which is initially over ionised. As the plasma expands and cools electrons recombine with ions and form the next lower ionisation stage. These ions tend to have their most highly excited levels populated to start with, but cascading processes within the energy levels soon populate the lower ones. The essence of these schemes is that the rate of expansion, and hence adiabatic cooling, of the prepared plasma is so rapid that the collisional radiative rates are inadequate to maintain equilibrium between upper and lower levels. A well documented case deals with population inversion between the $n=3$ and $n=2$ levels in hydrogenic ions leading to gain on the Balmer α transition. The desired conditions can be reached by heating a target of small dimensions in the expansion direction, e.g., a $5\,\mu$m diameter fibre, with a short laser pulse, e.g., a few picoseconds. The initial plasma is essentially composed of bare ions, a state which is reliably produced as it is akin to the closed shell ion configuration referred to earlier. Similar scenarios apply to 4–3 and 5–4 transitions ($\Delta n=1$) in lithium- and sodium-like ions recombining from initial helium- and neon-like hot plasmas, respectively. This class of laser, which has had some success at soft x-ray wavelengths, is broadly analogous to gas dynamic pumped carbon dioxide infrared lasers.

Collision pumped lasers also require the formation of a hot plasma with a predominant population of closed shell ions such as neon- and nickel-like. In the case of neon-like ions there is the possibility of exciting ground state (2p) electrons, through collisions with sufficiently energetic free electrons, to the 3p state. Collisional mixing in the $n=3$ levels and extremely rapid radiative decay from the 3s level back to the ground state then allows population inversion between the 3p and 3s sublevels; these schemes thus rely on $\Delta n=0$ lasing transitions. For nickel-like ions the population inversion is established between 4d and 4p sublevels. This scheme is unique to electron collision pumping as the primary pump mechanism is forbidden by photopumping. The collision pumped schemes have been by far the most successful to date and have their ancestry in the common argon ion (Ar II) optical laser whose 4p–4s lasing lines are pumped in a plasma discharge tube.

The geometry of an amplifying plasma is effectively determined by two things. Firstly, normal incidence multilayer mirror structures, although increasingly available with 10–30% reflectivities at wavelengths down to 4 nm, are prone to damage and layer diffusion if used too close to a laser plasma. Secondly, the terawatt pump powers needed to produce inversion cannot be sustained for more than a few nanoseconds in photopump or collisional pump schemes and the gain is transient on subnanosecond timescales in recombination schemes. The combination of these factors leads to most schemes involving only single pass operation where a plasma column, several hundred times longer than it is wide, is pumped to inversion. Spontaneously emitted photons of the laser transition are then preferentially amplified along the long axis giving an output beam from each end of the plasma. The divergence is given approximately by the ratio of the width to the length of the column. Such systems, in effect, produce amplified spontaneous emission. Their optical properties are intermediate to the incoherent output of a thermal laser plasma and the coherent output of a true laser in which the beam of radiation is amplified by multiple passes through the laser cavity. Even so, devices which work in the 3–40 nm wavelength range are commonly referred to as x-ray lasers.

As pump power technology, in the form of more powerful optical lasers, was maturing in the 1970s and early 1980s driven mainly by the inertial confinement fusion programmes, there was an increasing flow of reports of experimental progress in x-ray laser research based on laser plasmas. Most reports claimed evidence of population inversion either in spot focus plasmas or in short plasma columns (limited by pump power) where the gain-length product (αL) was so low that exponential growth was not unambiguous. Significant work was done on the recombination scheme using lasers to heat carbon fibres to provide gain on the C VI line at 18.2 nm, and aluminium slabs to provide gain on Al XI lines at 10.5 nm and 15.4 nm. Work in the late 1970s on collision pumped neon-like calcium plasmas suggested amplification at about 60 nm. In the early 1980s there were reports of x-ray laser emission from metre long, low density metal rods pumped by nuclear bombs in underground explosions in Nevada, rather than by laser plasmas. However, these also have a legacy of uncertainty in what was actually achieved, although they influenced the 'Star Wars' programme.

The demonstration of unequivocal evidence of amplification at a wavelength of 21 nm in a laser plasma, at Lawrence Livermore National Laboratory, USA, in the mid 1980s, triggered renewed interest in x-ray lasers which has been sustained worldwide ever since. The target used was a thin plastic foil coated with selenium and pumped at about 5×10^{13} W cm^{-2} either by a single laser beam from one side or by two beams from opposing directions. Initial experiments used about 1 kJ of laser energy at $\lambda = 0.53$ μm in 450 ps pulses to heat plasmas up to 2 cm long in a 200 μm wide line focus. Gain coefficients on the strongest, exponentially growing emission lines from 3p–3s ($\Delta n = 0$) transitions in neon-like selenium were about 5 cm^{-1}. From this starting point many other workers, mainly in the USA, UK, Japan, France, Germany and China, have made rapid progress in developing this type of collision pumped soft x-ray laser. Lasing action has been demonstrated for neon-like ions of the elements from titanium to silver, corresponding to roughly 100 discreet lasing lines with wavelengths in the range 10–45 nm.

Development of Collision Pumped X-Ray Lasers

Reports of neon-like lasing were quickly followed by similar experiments with higher atomic number targets used to demonstrate lasing action on 4d–4p (also $\Delta n = 0$) transitions of nickel-like ions. The first study used europium foil targets irradiated at 7×10^{13} W cm^{-2} (0.53 μm and 1 ns) and a gain coefficient $\alpha \approx 1$ cm^{-1} was measured on lines at $\lambda \approx 7$ nm. Many elements have now been studied between niobium and gold with nickel-like ion lasing wavelengths in the range 3.5–20 nm. The shorter wavelength regime is currently limited to researchers with access to the most powerful lasers available. The peak gain-length products achieved are about eight, equivalent to an amplification of about 3000. Although the nickel-like systems have the advantage of shortest wavelength, for a given pump power, it has been the neon-like ions that have been used for the systematic development of collision pumped x-ray lasers.

Following early work at the Naval Research Laboratory (USA), it was shown at the Central Laser Facility (UK) that germanium slab targets (or micrometre thick germanium layers on a substrate) heated by 1.06 μm laser beams lased very well at irradiances of about 10^{13} W cm^{-2} from about 1 ns pulses. This was important because such targets are less complex to fabricate and handle than the original foil targets. Subsequently, germanium has been used as a standard material in many laboratories. In common with most other similar ions, neon-like germanium lases on at least 5 transitions due to level splitting within the 3p and 3s sublevels. For germanium, the wavelengths of the strongest lines are 19.6, 24.7, 23.2, 23.6 and 28.7 nm. Initial experiments routinely observed brightest emission on the lines at 23.2 and 23.6 nm whereas theory predicted highest gain on the line at 19.6 nm.

The electron density and temperature are not uniform throughout the plasma. The volume in the germanium plasma which exhibits gain has an electron density of about 5×10^{20} cm^{-3} and a temperature of about 800 eV. The shorter the operating x-ray laser wavelength, the higher are the required density and temperature. Thus the shortest wavelength devices are pumped best with short wavelength optical lasers which can penetrate and deposit energy to higher electron density. However, since the transverse laser plasma dimensions are small high densities are associated with high density gradients. This leads to a serious problem since x-rays travelling along the plasma axis tend to move away from regions of high electron density in much the same way as radio waves bounce off the ionosphere. This is known as refraction and causes the amplifying beam to move out of regions of high gain after propagation over a characteristic distance known as the refraction length, typically about 2 cm. Since these x-ray lasers are single pass devices it is desirable to achieve as high a gain-length product as possible, limited only by the pump power. It was for this reason that the original thin foil design was used, the rapidly heated foil burning through during the drive pulse, exploding and, later in the expansion, providing a symmetric density profile with optimum magnitude and shallow gradients.

At the Central Laser Facility a new design based on short slab targets in tandem was introduced. The amplifying beam could zigzag along the axis to compensate for refraction, as illustrated in figure 10. These devices, known as double target amplifiers, have been very successful and widely used. Typically, output beams have divergences of several milliradians and are deflected off the main plasma axis by about 10 mrad.

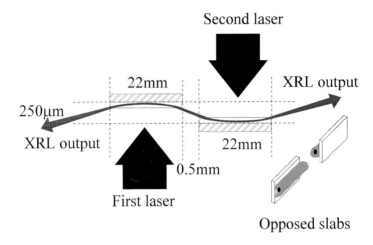

Figure 10. Schematic of a double slab target arrangement used to compensate for refraction effects in x-ray laser plasma amplifiers.

The next stage in x-ray laser development was to couple the output beam from one amplifier into another, not simply by proximity coupling as in the double target case but by relaying over a long distance as in multistage optical amplifiers. This has the potential of producing a much more coherent beam. Such a geometry is depicted in figure 11 where a double target is used as a beam 'injector' for the final amplifier. Coupling is achieved by imaging with an x-ray multilayer mirror and timing of the optical pump beams is synchronised to the propagation of the x-ray laser beam.

Amplification of a laser beam proceeds exponentially so long as the photon flux of the beam does not significantly deplete the population inversion on time scales comparable to the upper state lifetime. If this happens the gain is said to be saturated and the output beam intensity thereafter grows approximately linearly with amplifier length. Typically, saturated output is achieved in x-ray lasers

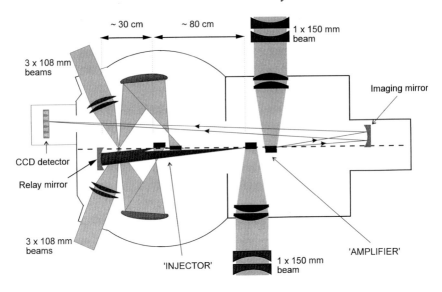

Figure 11. The x-ray laser injector/amplifier architecture used at the Central Laser Facility, UK. Targets are irradiated by the VULCAN neodymium-glass laser.

when the gain-length product of the amplifier exceeds 16. This has been achieved on several neon-like transitions and an example is shown in figure 12 for the 23.2 and 23.6 nm lines of germanium. The linear part of the growth curve indicates exponential growth (the intensity scale is logarithmic). Here the multilayer mirror coupling shown in figure 11 was modified to double pass the injector stage to achieve the highest possible gain-length product from a pumped double target. The brightest output reported to date is from a saturated yttrium x-ray laser at 15.5 nm where 40 MW output in 200 ps pulses was achieved. This represents an output of about 8 mJ ($\approx 6 \times 10^{14}$ photons) in a single pulse.

   With the ability to generate well defined beams of high brightness came the need to characterise and manipulate them, to allow their use in potential applications. The geometry of figure 11 can be modified so that the injector output beam is reflected from a multilayer mirror at 45°, the Brewster angle for x-rays. The normal incidence coupling mirror returns it for a second reflection and then to the final amplifier. This produces a linearly polarised beam at

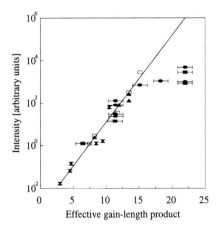

Figure 12. The exponential growth of x-ray laser lines at $\lambda=23.4$ nm up to saturation at megawatts power levels. Different symbols refer to different experiments.

$\lambda \approx 23.4$ nm using a germanium target. The 'footprint' of a 98% polarised beam is shown in figure 13(a); the beam divergence was 5 mrad and the missing parts of the image are due to thin crosswires used for alignment purposes.

Germanium systems which have reached saturation can have a high degree of spatial coherence. Significant effort has been devoted to quantifying this parameter, usually by placing obstacles (e.g., wire, slit and pinhole arrays) in the beams and observing the visibility of resulting interference patterns, such as that shown in figure 13(b). To obtain this the output of a germanium x-ray laser was incident on a Young's double slit with slit separation 100 μm and slit width 20 μm. The slits were overlaid with a ladder structure with about 300 lines per millimetre acting as a transmission diffraction grating so that the interference pattern for each lasing line was spectrally separated. The wavefront was sampled about 0.5 m from the output of the final amplifier in figure 11 and the visibility (contrast) seen in the brightest lines (the unresolved pair centred at $\lambda \approx 23.4$ nm) indicates good spatial coherence over scale lengths of ≥ 100 μm across the x-ray laser beam.

The longitudinal (or temporal) coherence of x-ray lasers has also been estimated indirectly using high resolution spectrometers to

a)   b)

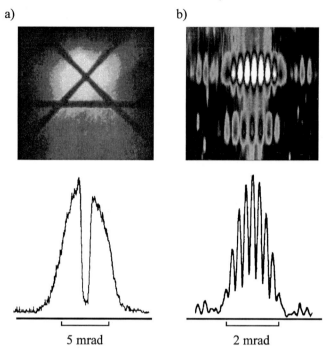

5 mrad   2 mrad

Figure 13. (a) A beam footprint of a linearly polarised germanium x-ray laser beam showing 5 mrad divergence; the beam shows the shadow of alignment crosswires. (b) The diffraction patterns and interference fringes observed when germanium x-ray laser lines are incident on a double slit structure. Fringes from the bright beams at $\lambda \approx 23.4$ nm are spectrally resolved from the fringes due to the weaker $\lambda = 24.7$ nm beam.

measure the spectral linewidths. As expected, linewidths narrow as saturation is approached in high gain-length situations and, typically, the longitudinal coherence is estimated to be a few hundred micrometres. This is a measure of the length of each x-ray wave packet along which there is a fixed phase relationship in the oscillation of the electric field. The longitudinal coherence length is a critical parameter for applications such as holography.

Many aspects of collision pumped x-ray lasers have been studied in some detail and the theory of these devices has been developed to a high degree of sophistication. However, a basic practical problem remains in that they are inefficient devices. Typically, saturated devices operating near $\lambda = 20$ nm require an investment of kilojoules of energy for a return of millijoules. This energy efficiency of about 10^{-6} (ignoring the efficiency of the optical pump laser itself) is partly due to the poor quantum efficiency of $\Delta n=0$ devices. To improve this situation there has recently been intense activity in trying to find ways to enhance the efficiency. New techniques which have all helped include the use of prepulses, pulse trains, travelling wave pumps and curved targets.

In the prepulse approach two pulses less than 1 ns long, separated by several nanoseconds, are focused onto the same target stripe. The first pulse intensity on the target is typically <1% of the second pulse and its function is to prepare a large volume, relatively cool plasma with which the main pulse interacts to produce gain. This has been shown to be very effective in producing much enhanced brightness on the neon-like transitions, usually known as $J=0-1$, which previously had been relatively weak compared to theoretical predictions. The improvement stems from the gain on these lines being highest in regions of the plasma where the electron density is highest, but normally the steepest density gradients also occur here and the observed growth of the lines is severely hampered by refraction. A preformed plasma allows heating and collisional pumping in gain zones of larger volume and shallower gradients. Output intensities can be enhanced by factors of 40 and pulse durations are typically about 10% of the second pulse, so that for optimising the efficiency only a short second pulse at the same intensity (i.e., less energy) is needed. This behaviour is peculiar to the $J=0-1$ lines as lasing on other lines lasts for a large fraction of the drive pulses, even up to several nanoseconds, and is independent of whether a prepulse is present.

Pulse train pumping has a similar philosophy to the prepulse case but now a series of two or more short pulses (≈ 100 ps) separated from each other by about 400 ps and of comparable intensity are used. The aim is to ionise a large volume uniform plasma with the first pulse and to rapidly heat this plasma with the second and subsequent pulses. The effect is demonstrated in figure 14 where the relative strengths of the germanium lines change dramatically going from the standard single nanosecond pulse drive to a double 100 ps pulse drive for similar slab targets and peak intensities.

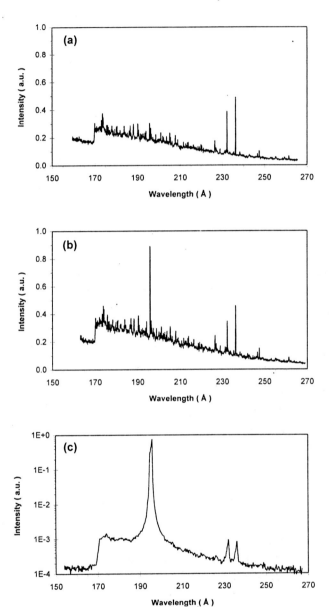

Figure 14. (a) Germanium x-ray laser lines at λ=19.6 nm, 23.2 nm and 23.6 nm from a 20 mm slab irradiated with an 800 ps pulse at about 1.5×10^{13} W cm^{-2}. (b) As in (a) but with a 2% prepulse 3.2 ns earlier. (c) A double 14 mm slab target (approximately half of one target was not heated well) irradiated at about 3×10^{13} W cm^{-2} by each of two 100 ps pulses separated by 400 ps. Note the increasing relative brightness of the 19.6 nm line and the logarithmic scale in (c).

The use of pulse trains leads to shorter x-ray laser pulses and this can also lead to inefficiency as the duration of the gain phase should be longer than the propagation time of a photon along the amplifier axis. In this case it becomes necessary to use travelling wave pumping whereby the phase front of the drive laser is tilted at 45° to its propagation direction using, for example, diffraction from an optical grating. In this way it is possible to pump the plasma in step with the arrival time of amplifying photons. This has been demonstrated at Lawrence Livermore National Laboratory with foil and slab targets giving output pulses as short as 50 ps and in a preferred direction. Pulse trains have also been used in Japan to pump nickel-like lines in the range 6–8 nm with observed gain-length products of about seven. These experiments also employed targets, cylindrically curved with radii of curvature about 2 m along the line focus, which helped to keep refracting photons in the high gain zone for as long as possible. This represents a first step in using waveguiding techniques to control the three-dimensional wave propagation through these inhomogeneous amplifiers.

A useful parameter in quantifying the efficiency of a scheme is the figure of merit, $\alpha L/E$, which gives the gain-length product that can be pumped practically for a given laser energy. An outline summary is shown in figure 15 of achievements to date using various target and pump conditions. The research goal can be defined as the intersection region of the line for saturated output with a kilojoule pump energy and the line for x-ray laser operation in the water window below 4.4 nm. Since 1985 there has been steady progress in this direction with collision pumped x-ray lasers.

Other Approaches to X-Ray Lasing

Although the collision pumped x-ray lasers described here have been the most successful to date there have been parallel efforts to develop new approaches and other schemes. In particular, recombination systems offer higher quantum efficiency, because $\Delta n = 1$, and recent experiments using 20 J/2 ps drive pulses from chirped pulse amplification neodymium-glass lasers have shown gain-length products $\alpha L \approx 6$ with $\alpha \approx 12$ cm^{-1}. Preliminary injection seeding experiments, where a gallium plasma amplifier was seeded with a coherent

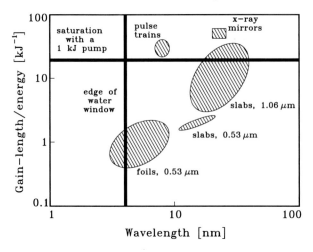

Figure 15. An indication of the gain-length production efficiencies achieved in various schemes of collision pumped x-ray lasers.

source of 25.1 nm photons, have shown amplification. The injected beam was produced through high harmonic generation where a 1.3 ps, 0.53 μm laser pulse (frequency doubled chirped pulse amplified beam) was focused into a helium gas jet to form multiple harmonics in the nonlinear interaction. Amplification of the resonant 21st harmonic was observed, giving an ultrashort x-ray laser pulse.

Developments in the near future could include nonlinear optical four wave mixing techniques to combine photons from x-ray lasers and optical lasers in a plasma environment. This could lead to doubled x-ray laser photon energy or to a degree of tunability for the x-ray laser. Current efforts relate to the use of high repetition rate table-top lasers with femtosecond pulses to pump inversions through recombination after optical field ionisation, creation of inner shell vacancies through ultrashort thermal x-ray flashes and collisional excitation. Such developments are seen to be crucial in the future role of x-ray lasers as widespread tools for applications.

Applications of X-Ray Lasers

As was the case for optical lasers the x-ray laser has not immediately found a wide range of applications. This is partly to do with accessibility to the large installations currently needed to provide the extremely bright beams and partly to do with the rather long wavelengths demonstrated so far for saturated output x-ray lasers. Nevertheless, many potential applications have been identified and preliminary demonstrations made in some cases. Meanwhile, laser developers are striving to provide better quality beams at shorter wavelengths and small laboratory scale drivers.

The characteristics of an x-ray laser beam determine its potential for a range of applications. In particular saturated x-ray laser pulses are extremely bright, monochromatic, quasi coherent and of short duration. At discrete wavelengths they are by far the brightest terrestrial sources available. Their capability for single shot contact microscopy has been demonstrated by imaging rat sperm cells at $\lambda=4.5$ nm with 50 nm resolution and test binary structures at $\lambda \approx 23.4$ nm with 150 nm resolution. A multilayer coated condenser, e.g., a Schwarzschild microscope objective or a concave mirror, is used to collect the x-ray laser beam and illuminate the sample which is then imaged at high magnification with a Fresnel zone plate.

Demonstrations of Gabor holography have been made where test objects in the x-ray laser beam scatter photons which then interfere with the unscattered reference beam. It is expected that Fourier transform holography will be demonstrated soon where the scattered radiation interferes with a spherically expanding reference wave enabling higher resolution in the magnified image. These are steps towards what has long been considered one of the primary applications of an x-ray laser, i.e., to make holographic (or three-dimensional) images in the water window of macromolecular biological samples *in vivo*. In principle, this is possible as the x-ray laser pulse is so bright and short that the exposure necessary to capture an image can be delivered in a single shot and before the sample is destroyed by the high intensity.

Applications in materials science, not yet demonstrated clearly, take advantage of the x-ray laser monochromaticity and include photoelectron spectroscopy and microscopy. Irradiation of a sample surface (e.g., a semiconductor device) causes a clean spectrum of ejected electron energies which reflect the chemistry and core level electronic structures of the constituents near the material surface. Visible lasers have been useful in the study of atomic structure and dynamics associated with valence electrons, and x-ray lasers should play a similar role in fundamental studies of inner shell atomic physics.

Some of the first practical applications of optical lasers were as probe beams for precise scientific measurement and x-ray lasers are following in their footsteps. Because of their association with large fusion laboratories it is not surprising that the first scientific applications relate to the diagnosis of dense plasmas. X-ray laser beams, with their short wavelength, high brightness and short duration, are able to propagate through long pathlengths of relatively high density transient plasma without refraction.

The yttrium laser at $\lambda=15.5$ nm has been used at Lawrence Livermore National Laboratory to image electron density structures in laser plasma plumes in exploding foils with 1 μm resolution. The germanium laser at $\lambda\approx23.4$ nm has been used at the Central Laser Facility to sidelight thin plastic foils which were imaged during their flight after acceleration by a 1 ns driving laser. An example is shown in figure 16 where a lollipop target (a thin disc supported on a thin

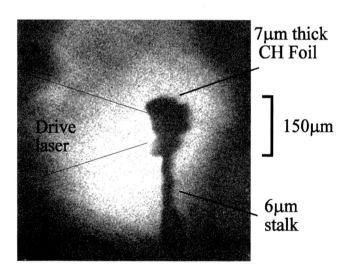

Figure 16. An edge-on image of a polystyrene foil target ablatively driven and accelerated by a 1 ns laser beam focused to about 6×10^{14} W cm^{-2}. Illumination is with a 300 ps germanium x-ray laser, freezing the material motion about 1.5 ns after the peak of the driving pulse. Magnification with the multilayer mirror imaging system is 4.2× and spatial resolution is 2.5 μm.

carbon fibre) replaced the final amplifier of figure 11 and was shot with the beam that normally pumps the amplifier. Experiments where the x-ray laser is incident from approximately the same direction as the laser beam which accelerates the foil target allow high resolution two-dimensional in-flight imaging of the target's mass distribution and hence can be a powerful diagnostic of hydro-dynamic instabilities.

Moiré fringe deflectometry, which is a technique for measuring small angular deviations induced in the direction of a radiation beam, has been used at Lawrence Livermore National Laboratory using an yttrium x-ray laser to measure density gradients in a laser plasma formed by irradiating a thick plastic foil. Recently, a Mach-Zehnder interferometer was built from multilayer mirrors and beam splitters, which are multilayer mirror structures on $100\,nm$ thick silicon nitride substrates and are partially transmitting and partially reflecting. This has been used to measure electron densities up to $3\times10^{21}\,cm^{-3}$ in a laser heated polystyrene foil. The yttrium x-ray laser beam was split into two and then recombined to give a two-dimensional reference pattern of parallel interference fringes. One beam, which penetrates a laser plasma, can have its optical path length increased due to the presence of free plasma electrons and the effect is quantified by measuring fringe shifts in the reference pattern.

Acknowledgements
All the data shown was recorded using lasers at Queen's University, Belfast or at the Central Laser Facility at the Rutherford Appleton Laboratory. The expert help of many colleagues and students was invaluable in executing experiments and preparing illustrations. In particular, I thank G F Cairns, S J Davidson, A G MacPhee, D Neely, M Pringle, G Slark and I C E Turcu.

Bibliography
Since this article was intended for a general audience, I have taken the liberty not to detail the references for the many technical papers which have been published by many workers in these areas. Instead, for the interested reader, I include a limited bibliography of useful

start points to delve deeper into the general subject area. Comprehensive references to most of the material included here (and much more) can be found within these texts.

R A Cairns & J J Sanderson (eds) "Laser-Plasma Interactions" Edinburgh: SUSSP Publications (1980)

R A Cairns (ed) "Laser-Plasma Interactions 2" Edinburgh: SUSSP Publications (1983)

M B Hooper (ed) "Laser-Plasma Interactions 3" Edinburgh: SUSSP Publications (1986)

M B Hooper (ed) "Laser-Plasma Interactions 4" Edinburgh: SUSSP Publications (1988)

M B Hooper (ed) "Laser-Plasma Interactions 5" Edinburgh: SUSSP Publications (1995)

R C Elton "X-Ray Lasers" San Diego: Academic Press (1990)

P Jaeglé & A Sureau (eds) "Proc. 1st International Coll. on X-Ray Lasers" *J. de Physique, Colloque* **C6** (1986)

G J Tallents (ed) "X-Ray Lasers 1990" *IOPP Conf. Series* **116** (1991)

E E Fill (ed) "X-Ray Lasers 1992" *IOPP Conf. Series* **125** (1992)

D C Eder & D L Matthews (eds) "X-Ray Lasers 1994" *AIP Conf. Proc.* **332** (1994)

How Lasers Generate Bright Sources of X-Rays

Mike Key
Rutherford Appleton Laboratory and Oxford University

P rogress in the uses of electromagnetic radiation has frequently depended on the development of more intense sources. The most dramatic example of enhancement of intensity came in the 1960s with the development of lasers. Light amplification by stimulated emission of radiation bypasses the Planckian thermal intensity limit and intense laser beams quickly revolutionised the uses of electromagnetic radiation over a wide wavelength range, from the infrared to the ultraviolet.

Lasers did not, however, have an immediate impact on x-ray sources. Einstein's fundamental relationships governing spontaneous and stimulated emission probabilities dictate that spontaneous emission is much more probable than stimulated emission at high frequencies. This means that to create x-ray lasers requires extremely high power input to sustain the excited atomic states, from which the laser action occurs, against spontaneous decay[1]. For this reason, there were initially no x-ray lasers.

The development of x-ray sources occurred much earlier as evidenced by the present centenary celebration of the discovery by Röntgen. It proceeded on a course dominated by the use of beams of accelerated electrons. In the classical x-ray tube accelerated electrons strike a solid anode where abrupt deceleration produces 'bremsstrahlung' (which translates to 'braking radiation'), giving a continuous spectrum of x-rays at photon energies up to the full energy of the electrons. This is supplemented by emission of characteristic x-ray lines. The discrete line emission occurs when an inner shell vacancy, produced by collisional ejection of an atomic electron, is filled by an outer shell electron with the emission of an x-ray photon. Brighter x-ray tubes require more intense electron beams and the brightest conventional x-ray sources were produced with the highest current density of electrons in tightly focused beams. There is a limit to the intensity of the x-ray emission resulting from the limit on the electron beam current density due to heating and vaporisation of the anode.

A breakthrough in intensity of x-ray sources occurred, at about the same time as the development of lasers, due to progress in the technique of x-ray generation using synchrotron radiation. Here, the accelerated motion of electrons needed to produce electromagnetic radiation comes not from an abrupt collision with an anode but from the gentler deflection of an electron beam in vacuum by a magnetic field. Electron beams carrying much higher power per unit area can be produced in circular electron beam storage rings coupled to linear accelerators. The beams are stored at a electron energy in the gigavolt range, much above the tens of kilovolt energy in a conventional x-ray tube. At such high energies the mutual Coulombic repulsion between the negative charges of the electrons is less effective in defocusing the beam and very high power density can be obtained. The less violent acceleration in the magnetic fields of bending magnets, wigglers and undulators applied to the electron beam in a storage ring produces, from gigavolt electrons, x-ray photons with energies from a fraction of a kilovolt to a few tens of kilovolts. Synchrotron radiation sources greatly enhanced the available intensities of x-rays and have become major tools of scientific research as discussed in this volume by Ian Munro.

In recent years the development of optical lasers has also begun to impact on x-ray sources. Motivation to generate extreme power and focused power density in optical laser beams came from the objective of igniting thermonuclear fusion[2]. The new fusion laser technology opened the way to new kinds of x-ray sources which can be much brighter than even the synchrotron radiation sources. Conventional x-ray tubes and synchrotron radiation sources both emit essentially continuously, although in the case of synchrotron radiation the continuous power is modulated into short bursts at megahertz frequency. Laser driven sources, in contrast, emit powerful single pulses. This key difference makes them complementary to the continuous sources and of interest where short duration single pulse exposures are needed to freeze quickly changing events.

The fusion laser technology led first to powerful sources of thermally emitted soft x-rays from hot plasmas and then to soft x-ray lasers based on radiative transitions between electronic states of highly ionised ions created in hot laser produced plasmas. For an x-ray laser the plasma is produced via a narrow line focus a few centimetres long on a solid target. Inversion of population density between particular electronic states of the ions is created in various ways, causing amplification and exponential build up of spontaneously emitted radiation along the plasma column. As no laser resonators have yet been developed, extremely high single or double pass exponential gain (typically by a factor of more than 10^6) is used to reach the saturation level where stimulated emission dominates over spontaneous emission These sources are discussed in more detail in the article in this volume by Ciaran Lewis.

A more recent development has been an extension of the technology of lasers to produce ultrashort pulses at very high powers. New solid state laser materials and novel optical techniques have allowed the routine generation of pulse lengths in the range of a few tens to a few hundreds of femtoseconds[3]. Used together with ingenious optical techniques, these new ultrashort pulses have greatly increased the power and focused intensity of optical lasers. The result has been further progress towards novel x-ray sources and it is this area which is discussed in this chapter. First, basic ideas of intensity and the characteristics of the new optical frequency lasers will be described briefly. Application to the production of uniquely intense and coherent sources of XUV emission through the generation of high harmonics of the optical frequency will then be outlined. The impact of the new optical lasers on the development of x-ray lasers from plasmas is then discussed, together with the interesting process of production of exceptionally intense electron beams accelerated by laser light which gives a novel variation on traditional anode x-ray sources. Finally, the classical subject of radiography with x-rays will be revisited with a description of the use of soft x-ray lasers for highly specialized radiography of matter accelerated to extremely high speeds in laser fusion research.

Intensity in Radiation Sources

The intensity I of a radiation source is defined as the power per unit area per unit solid angle and is the key parameter determining the effectiveness of the source for most applications. A convenient example is the exposure level in a radiograph of specified spatial resolution R recorded with a distance F between the object and the film. If the source is of diameter D located at a distance $S \gg F$ from the object, then the spatial resolution of the radiograph is $(F/S)D$. The exposure level E, or power per unit area, on the film is IR^2/F^2, which is proportional to the intensity of the source. More generally, the signal level in most measurements is proportional to the intensity which is an invariant parameter in lossless optical focusing systems.

New High Power and Ultrashort Pulse Lasers

Recent progress in laser technology has led to exceptionally high focused power density (power per unit area)[3]. This is conveniently illustrated by three different laser systems, all developed at the Rutherford Appleton Laboratory, which have reached the highest focused power densities obtained worldwide.

First, it is instructive to consider the factors limiting focused power density. The previous discussion about intensity of radiation sources is applicable to lasers with the important consideration that laser light is coherent and laser beam propagation is governed by diffraction. We need to maximise the focused power density P_f of the laser beam; the essential factors determining this can be seen in the expression $P_f = P_b w_c^2 / \lambda^2 f_d^2 = I/\lambda^2 f_d^2$ where P_b is the power density in an unfocused laser beam, w_c is the width over which the beam is coherent (equal to the full width for a diffraction limited beam), f_d is the focal length to diameter ratio of the focusing optic and λ is the laser wavelength. The beam intensity is the key to recent progress to higher focused power density. The previous limit on P_b was from break up of the coherent wavefront by exponential growth of small scale intensity perturbations (bright spots) through self focusing. In this process the beam refracts towards the centres of bright spots due to the increase of refractive index, which becomes significant at very high power density. Self focusing generally restricts power density to a few gigawatts per square centimetre in conventional solid state lasers operating at their self focusing limit.

A dramatic increase has been obtained with the new technique of chirped pulse amplification (CPA)[3]. Stable generation of ultrashort pulses in the range 20 fs–1 ps from new solid state laser pulse generators, notably the titanium sapphire laser, has been a vital factor in this development. The ultrashort pulses have a wide frequency bandwidth and, by reflection from a pair of diffraction gratings, different optical paths are created for the short and long wavelength components of the pulse spectrum. The pulse length is thus stretched and chirped, its frequency changing linearly in time. The broad bandwidth long duration stretched pulse is amplified and reaches a high energy per unit area without exceeding the intensity limit for self focusing. It is then compressed to short duration and much higher power density using a second pair of large aperture diffraction gratings in vacuum, where there is no self focusing. The only limit on P_b is damage to the reflecting surfaces of the gratings.

Application of this scheme to the Vulcan neodymium glass laser at the Rutherford Appleton Laboratory has given a hundredfold increase in its single beam power output[4]. The power from a 15 cm diameter amplifier now reaches 35 TW at the fundamental wavelength of 1.05 µm in a pulse of 0.6 ps minimum duration. The beam is coherent over about 25% of its aperture. An $f/3$ off axis parabolic mirror focuses the beam to spot of 20 µm diameter at an intensity of $10^{19}\,W\,cm^{-2}$. The upper limit is set by the $0.2\,J\,cm^{-2}$ damage threshold of the 15×30 cm reflection gratings, which are used to recompress the pulse from its stretched duration of 200 ps. Figure 1 shows the compression and beam focusing system.

The same chirped pulse amplification principle has been applied also to a 27 cm aperture KrF laser, Sprite, at the Rutherford Appleton Laboratory[5]. In this case the laser operates at a lower energy per unit area and the pulse is stretched to 15 ps before amplification and recompression to the 300 fs bandwidth limited minimum pulse duration of the KrF laser. The beam is telescoped down onto a pair of compression gratings of 10 cm aperture and their

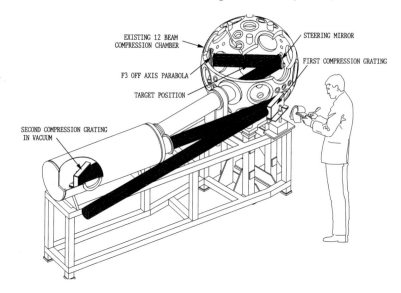

Figure 1. The experimental apparatus for pulse compression and beam focusing at the Vulcan laser.

damage threshold limits the power to 1 TW. With an $f/3$ parabolic reflector at the ultraviolet wavelength of 248 nm and with three times diffraction limited beam quality, the focal spot size is only 3 mm and the intensity is almost equal to that of Vulcan.

Similar focused intensity is obtained with the Sprite laser by an entirely different and unique method generating longer (10–20 ps) pulses which are of interest for applications needing more pulse energy. The laser is configured as a KrF laser pumped Raman laser using a series of beam combining Raman amplifiers[4], as illustrated in figure 2. The nonlinear limit on beam intensity is circumvented because the Raman amplifiers are simply gas cells (typically 1 m long and filled with methane at 1 bar) and they have only millimetre thick fused silica windows. Energy is extracted from the 27 cm diameter KrF amplifier by angularly separated beams with relative delays of the pulse in each beam enabling the stored energy in the laser to be replenished in the 5 ns interval between pulses. The individual pulses have power levels below the self focusing limit for duration >10 ps. The KrF laser beam paths are adjusted to make the pulses arrive synchronously at the Raman amplifiers. A single beam at the Raman shifted wavelength for methane (268 nm) is amplified and efficiently extracts the energy of multiple pump beams in a beam of near diffraction limited quality. The Raman mode of operation thus also gives focused intensity up to $10^{19}\,\mathrm{W\,cm^{-2}}$.

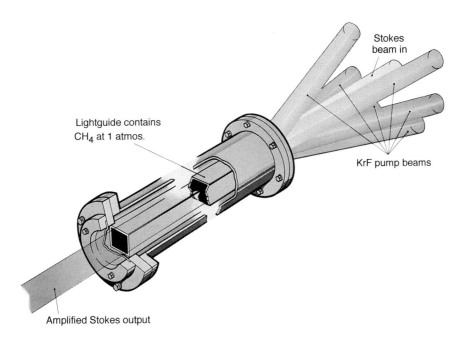

Figure 2. A beam combining KrF laser pumped Raman amplifier showing the gas cell, angled pump beams and axial Stokes frequency Raman laser beam. (Courtesy of Rutherford Appleton Laboratory)

Many laser systems worldwide are now operating with focused power density above $10^{18}\,\mathrm{W\,cm^{-2}}$, most using the CPA technique. Systems of 'table top terawatt' scale have made such high power densities widely available[3] and lasers of this class are opening up significant new opportunities for the more widespread development and use of novel sources of XUV and soft x-ray radiation.

XUV Harmonic Generation
A most dramatic effect of high power density is the generation of extremely high harmonics of the optical frequency which provide uniquely short duration, powerful and coherent sources at XUV

wavelengths[3]. A simple physical picture explains the essential features of this process. The strong electric field of the laser radiation distorts the potential wells binding electrons to atoms and electrons escape by quantum mechanical tunnelling. The process is termed optical field ionisation. The freed electrons begin an oscillation in the electric field of the laser radiation which returns them to the vicinity of the ion after one half cycle. There is then a finite probability of a collision in which the electron recombines with the ion giving up both its oscillatory kinetic energy and its ionisation energy to an emitted photon with many times the energy of the optical laser photon. The process is coherent with the optical laser and the harmonic waves can build up to large intensities. A full spectrum of the odd harmonics of the laser frequency is generated up to a high frequency limit where the photon energy is equal to the sum of the oscillation energy (which scales as $P\lambda^2$) and the maximum ionisation energy (which is a function of the power P).

Until recently the highest harmonic photon energies were obtained by maximising $P\lambda^2$ using a long wavelength laser such as titanium sapphire (74.4 nm, 109th harmonic) and neodymium glass (73.6 nm, 143rd harmonic)[3]. Neutral atoms were used in order to avoid loss of harmonic conversion efficiency due to loss of phase

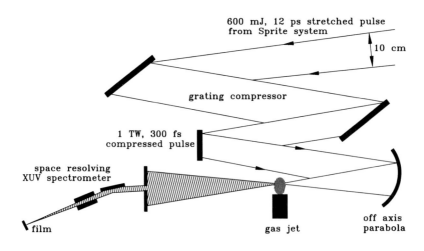

Figure 3. Apparatus for XUV harmonic generation with the Sprite chirped pulse amplified laser beam.

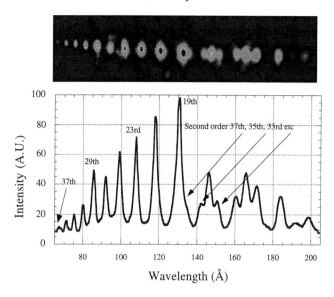

Figure 4. A spectrum of XUV harmonics recorded with the Sprite chirped pulse amplification laser.

correlation between the optical and harmonic waves. This occurs through the strong frequency dispersion of the refractive index of free electrons, but use of neutral atoms also limits the peak power density to $<10^{16}\,\mathrm{W\,cm^{-2}}$ because at higher levels the majority of neutral atoms are field ionised before the laser pulse builds up to its peak power. The free electron dephasing effect is less serious for a lasers of shorter wavelength and very recent experiments have used the ultraviolet CPA laser beam from Sprite, described previously, to increase the efficiency of generation and reduce the shortest observed harmonic wavelength to 6.72 nm, which is the 37th harmonic of the KrF laser beam[7]. Figure 3 shows the experimental set up and figure 4 the spectrum of the harmonics recorded by focusing the chirped pulse amplified KrF laser beam into a 1 mm wide pulsed gas jet of helium atoms. Optical field ionisation of the ion He^+, rather than neutral atoms, was the source of the harmonic emission. The intensity for this process $(5\times10^{17}\,\mathrm{W\,cm^{-2}})$ is still rather low compared to the maximum available from the laser and it is interesting to speculate how far this approach can be developed.

Soft X-Ray Lasers

Soft x-ray lasers operating at wavelengths of 4–40 nm with mega-watt pulse power and a fraction of a nanosecond pulse duration have now been demonstrated with plasmas of highly ionised ions as the laser media[8]. The plasmas are produced by heating solid targets with optical laser pulses of energy ranging from hundreds to thousands of joules in subnanosecond pulses as discussed by Ciaran Lewis. These lasers operate mainly by electron collisional excitation of highly ionised neon-like and nickel-like ions in hot plasmas.

An important objective of current research is to reduce the laser pulse energy needed to drive the x-ray lasers because at the moment the energy required means that the driver has to be an expensive large scale laser facility, which severely limits the exploitation of x-ray lasers by potential users. A convenient measure of the effectiveness of the driver laser is the ratio of the x-ray amplification exponent (the product of gain coefficient and length) to the drive energy, as illustrated in figure 5. The gain exponent should reach

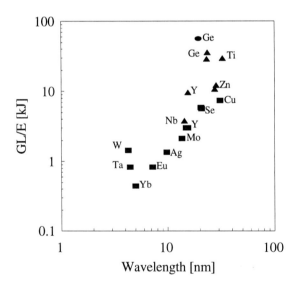

Figure 5. Efficiency of producing gain in XUV lasers. Squares denote exploding thin foil targets driven by green light, triangles are solid slabs driven by infrared light and the circle shows a solid slab with infrared drive and a small prepulse. Wavelengths below 10 nm are from nickel-like ions and above 10 nm are from neon-like ions.

15–20 in order to achieve saturated laser action, i.e., dominance of stimulated over spontaneous emission, in single pass amplification of spontaneous emission. Figure 5 shows how, for conventional x-ray lasers, this requires an increasing number of kilojoules of drive energy for wavelengths below 20 nm.

A dramatic reduction in driver energy has been demonstrated in recent work with the CPA beam of Vulcan. It had been shown several years ago that XUV laser action on the 18.2 nm Balmer α transition of the hydrogen-like ion of carbon could be obtained in a plasma of bare nuclei of carbon recombining rapidly to form hydrogen-like ions in the course of explosive adiabatic cooling. This was achieved by laser irradiation of a carbon fibre using six high energy beams from Vulcan operating at a pulse duration of 150 ps. Theoretical work had shown that much higher gain with lower driver energy could be obtained with the new short pulse CPA beam. Confirmation of this prediction has recently been obtained with a demonstration of amplification by a factor of 400 in only a 5 mm length of plasma using just 20 J in 2 ps to irradiate the 7 mm carbon fibre[9], as shown in figure 6. The improved C VI laser has

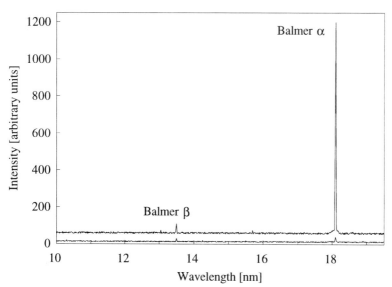

Figure 6. C VI laser spectra for plasma lengths of 1 mm (lower trace) and 5 mm (upper trace).

demonstrated the highest gain coefficient of any XUV laser. The efficiency of gain production is increased by a factor of more than an order of magnitude relative to conventional x-ray lasers of similar wavelength, as illustrated in figure 5.

The topicality of this field is further illustrated by a report, at a conference last week as I write this article, of greatly improved gain efficiency in an XUV laser operating at 34 nm in a collisionally excited plasma of neon-like ions of titanium. A conventional nano-second pulse irradiation at rather low intensity and only a few joules of energy was used to preform a plasma which was suddenly reheated by a 1 ps CPA laser burst of much higher intensity (with just two joules of energy). The increased rate of electron collisional excitation of the plasma ions boosted the laser gain coefficient to a value similar to that in the C VI laser. The gain efficiency illustrated in figure 5 is about 30 times higher than for conventionally excited titanium lasers[10].

The new class of low energy and short pulse driven x-ray lasers promises to make these novel devices more widely available. It can also be foreseen that, with the most powerful short pulse drivers, there will be progress to shorter x-ray wavelengths.

Laser Generated Electron Beams and Novel X-Ray Sources

Recent work with very high laser intensities, up to 10^{19} W cm^{-2}, on solid targets has shown that the laser beam can be quite efficiently converted to a flow into the target of energetic electrons. The laser energy is absorbed at the critical cutoff density for electromagnetic wave propagation (10^{21} electrons per cubic centimetre for a 1 μm laser wavelength) and transformed into energy of plasma electrons. These electrons have a mean energy of up to 200 keV (as can be seen by equating the absorbed energy flow of the laser at up to 10^{18} W cm^{-2} to that of electrons at the critical density travelling at the speed of light). The target material acts like a conventional anode and converts a fraction of the electron beam energy to x-rays[11]. The important point is that the electron beam intensity, at up to 10^{18} W cm^{-2}, is many orders of magnitude higher than in a conven-tional x-ray tube. Target vaporisation is avoided by the short pulse duration and the space-charge limit on beam intensity is avoided because the medium is an electrically neutral plasma. Table top terawatt lasers operating at tens of hertz repetition rate can produce hard x-ray bremsstrahlung and characteristic K_α pulses of exceptional brightness and are being evaluated, for example, as advanced tools for high resolution medical radiography[12].

Comparison of XUV and X-Ray Sources

In comparing x-ray sources it is usual to consider the spectral brilliance, which is the intensity per unit of frequency bandwidth. The XUV harmonic and soft x-ray laser sources give by far the highest peak brilliances in their region of the spectrum, generating between about 10^{21} and 10^{24} photons s^{-1} mm^{-2} mrad^{-2} in 0.1% bandwidth in very short pulses. They are restricted, however, to longer wavelengths than synchrotron radiation and anode sources, as shown in figure 7 where they are compared with other sources. The brightest synchrotron radiation sources (undulators) have about four orders of magnitude lower instantaneous brilliance and conventional x-ray tubes are ten orders of magnitude lower still. The harmonic sources have, typically, subpicosecond pulse duration; their coherence is high but their pulse energy is only in the nanojoule range. The XUV

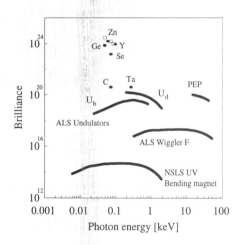

Figure 7. Instantaneous spectral brilliances (photons s^{-1} mm^{-2} mrad^{-2} in 0.1% bandwidth) of XUV and x-ray sources. XUV lasers are shown as filled circles with the element name, harmonics are open squares and synchrotron radiation sources have the machine names.

lasers emit single pulses of a fraction of a nanosecond duration with less coherence than the harmonic sources but much larger (milli-joule) energy. The undulator sources emit a continuous train of, typically, 50 ps pulses at several orders of magnitude lower brilliance per pulse but at 100 MHz repetition rate, and so have vastly greater average power.

An Application of XUV Lasers to Radiography

The interesting applications of the laser generated XUV and x-ray sources are those which demand high brilliance in single pulses. They do not compete with undulators in applications requiring high average power but can be used for single pulse applications impossible with undulators.

A good example of such an application which could be carried out only with an XUV laser is very recent and used the multi-megawatt power and subnanosecond pulse duration of a neon-like yttrium laser operating at a wavelength of 15.5 nm. The objective was to measure, by transmission radiography, variations of a few

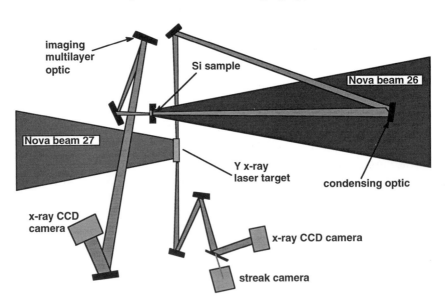

Figure 8. An experimental system for XUV laser radiography showing the path of the XUV laser beam and the high energy beams of the Nova laser driving the XUV laser and the silicon foil target.

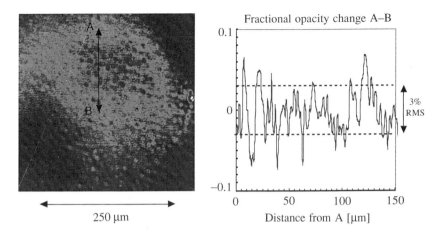

Figure 9. A radiographic image showing thickness variations induced by laser speckle in a laser irradiated thin foil target.

percent in the thickness of a foil target just at the moment when a laser driven shock wave had passed through it[13]. The experimental system is illustrated in figure 8. The context is laser fusion research, where it is required to achieve very uniform spherical implosions driven by intense laser light. The coherent speckle pattern associated with laser light causes small variations in the drive pressure which induce perturbations in the target on a spatial scale of a few microns. The test target probed with the yttrium laser was 3 µm thick and the induced thickness variations were measured after 200 ps of ultraviolet laser irradiation at $0.6 \times 10^{13}\,\mathrm{W\,cm^{-2}}$. These were conditions mimicking the first stage of a laser fusion implosion. The picture in figure 9 was obtained by projecting a radiographic image of the foil at high magnification onto an XUV sensitive CCD camera using a spherical multilayer mirror. It shows the pattern of thickness variations which are only about 150 nm in amplitude and with spatial scale down to a few micrometres. The very strong absorption of the XUV laser beam in the target gave high sensitivity of the transmission to small variations of thickness and its high brightness enabled the image to be recorded with 10 000 times attenuation in the target. Such a measurement could be made only with an XUV laser.

Conclusions

Production of extremes of power density using focused beams from new ultrashort pulse lasers is having significant impact on the generation of XUV and x-ray sources of exceptional brightness. New developments are occurring in both harmonic generation and x-ray laser action and this area of research is fast moving and can be expected to lead to still more intense and shorter wavelength laser generated x-ray sources.

Special emphasis on the development of high intensity lasers and on their scientific applications has been a feature of recent work at the Central Laser Facility (CLF) at the Rutherford Appleton Laboratory in the UK. Of particular importance for the laser developments has been a joint venture with UK groups from Southampton University, Imperial College and St Andrews University and also collaboration in KrF lasers with Professor Schäfer at the Max Planck Insitut für Biophysikalishechemie. Access to the lasers at the CLF, which are provided by the UK Engineering and Physical Sciences Research Council, is open to all UK University researchers and recently access for European teams has been provided in the European Commission Human Capital and Mobility programme.

Acknowledgments

I am indebted to many colleagues for help in connection with this paper and I would like to thank particularly Dan Kalantar at Lawrence Livermore National Laboratory and Justin Wark and Stephen Preston at Oxford University for supply of unpublished information.

References

(1) M H Key "Laboratory Production of X-Ray Lasers" *Nature* **316** 314 (1985)

(2) M H Key "Lasers Generate Plasma Power" *Physics World* p52 (August 1991)

(3) M D Perry and G Mourou "Terawatt to Petawatt Subpicosecond Lasers" *Science* **264** 917 (1994)

(4) C N Danson *et al* "High Contrast Multi-Terawatt Pulse Generation using Chirped Pulse Amplification on the Vulcan Laser Facility" *Opt. Comm.* **103** 392 (1993)

(5) I N Ross *et al* "A 1 TW KrF Laser Using Chirped Pulse Amplification" *Opt. Comm.* **109** 288 (1994)

(6) M J Shaw *et al* "Ultrahigh Brightness Laser Beams with Low Prepulse Obtained by Stimulated Raman Scattering" *Opt. Lett.* **18** 1320 (1993)

(7) S Preston *et al* "High Order Harmonics of 248.6 nm KrF Laser from Helium and Neon Ions" *Phys. Rev. A* **53** 31 (1996)

(8) B J MacGowan *et al* "Short Wavelength X-Ray Laser Research at the Lawrence Livermore National Laboratory" *Phys. Fluids* **B4** 2326 (1992)

(9) J Zhang *et al* "Demonstration of High Gain in a Recombination XUV Laser at 18.2 nm Driven by a 20 J, 2 ps Glass Laser" *Phys. Rev. Lett.* **74** 1335 1995

(10) P Nickles *et al* "An Efficient Short Pulse XUV Laser on Neon-like Titanium" *Proc SPIE* **2520** 373 (1995)

(11) J D Kmetec *et al* "MeV X-Ray Generation with a Femtosecond Laser" *Phys. Rev. Lett.* **68** 1527 (1992)

(12) S Svanberg *et al* "Lund High Power Laser Facility—Systems and First Results" *Physica Scripta* **49** 187 (1994)

(13) M H Key *et al* "Measurement by XUV Laser Radiography of Hydrodynamic Perturbations in Laser Accelerated Thin Foil Targets" *Proc SPIE* **2520** 282 (1995)

The Next Hundred Years

Andrew Miller
University of Stirling

P rediction of what will happen in basic research is impossible. This is true in principle, and has been shown many times in practice. As to the discovery of x-rays in 1895, Lord Kelvin has been quoted as saying that they would never amount to anything. More recently, I was told in the early 1980s, by several most eminent UK scientists, first that x-rays from synchrotron radiation would never be useful for diffraction studies, secondly and more specifically that synchrotron radiation would never be useful for protein crystallography and, thirdly that a bright x-ray beam would destroy biological samples and render experiments of no value. Within two or three years these predictions were obviously wrong. Prediction in basic research, even of one year ahead, is often impossible, especially by committees.

So why continue with this article? Can it be anything more than a comfortable entertainment to round off this excellent celebration of a hundred years of success for x-rays? Perhaps I should hazard a more general prediction—demonstrating how little some of us learn from failures in the past—by saying that, while I do not expect to be around myself in 2095 to assess the fate of my predictions, I would expect, given the advances in medical science, that some readers in their twenties (or younger!) might well be alive at that time. You can see that I have decided to proceed with caution.

A different prediction to what might happen in basic research is to consider what might happen in the development of the application of knowledge. The history of this activity is also replete with failures, most famously the view of Lord Rutherford that nuclear physics would have no applications. Despite this, it is certainly

X-RAYS: The First Hundred Years
Edited by Alan Michette and Sławka Pfauntsch © 1996 John Wiley & Sons Ltd

easier to extrapolate to possible applications of basic science than to predict the course of basic research, and one assumes that it is future applications that the UK government has in mind in the Technology Foresight Exercise. Yet it must be emphasised that reality is quite complex.

The well known Frascati definitions from 1963[1] attempted to distinguish between basic research, "…the search for new knowledge which does not necessarily have a technological objective, but which ultimately may very well influence technology…", and applied research, "…systematic activity directed to the advance of technology…". Such a neat distinction between basic and applied research is often not appropriate and, while helpful, does not encompass things as they really are. All medical, agricultural and engineering science is by no means always simply characterised as applied and much of what passes for basic research is far from creative. For example, routine repetition of some molecular biology or protein crystallography experiments may well add to the sum of human knowledge in the sense that natural history observations do. These experiments need a population of well trained, competent and disciplined scientists, but a Sanger, a Crick, a Perutz or a Tinbergen—to mention only some scientists who worked in the UK— have always been needed to make the creative advances in the basic sciences.

With this background I will make some guesses on how the future might go with x-rays.

X-Ray Sources

The possible futures of synchrotron and laser based sources have already been discussed in this volume. When the European Synchrotron Radiation Facility was being designed in 1986, Godfried Mullhaupt and his colleagues were already thinking about the next generation synchrotron source in terms of increased brilliance. They produced a design intended to decrease the emittance of the beam by another order of magnitude and noted that, for a wavelength of around 0.1 nm, the emittance would approach the diffraction limit of such a beam. We see, therefore, that while such an improved emittance can be expected, the laws of physics are such that further improvements will be constrained.

There are also designs for new sources tailored for specific purposes and some of these are being built. So called table-top synchrotrons are already available for lithography and hence applications to the microelectronics industry. Higher energy small synchrotrons have been envisaged which also use high field superconducting magnets. These might produce x-rays around the iodine or gadolinium absorption edges and hence provide the basis of hospital based coronary angiography diagnostic devices. They could also be used in the imaging of specific biological tissues by contrast variation.

More speculatively, one can think of the exploitation of nanotechnology, perhaps using particle or electromagnetic beams, to fashion crystals or molecular assemblies which might channel accelerated electron beams to produce nano-sources of x-rays.

Detectors

At present, the performances of x-ray sources have outstripped the ability of available detectors to cope with them. It is likely that progress will take place in the development of arrays of single pixel photon counting detectors to cope with the very high count rates in specific areas of scattering patterns.

It would also be helpful if detectors containing small pixels with energy resolution were available. This could improve the safety and efficiency of, for example, coronary angiography by improving the amount of information obtainable for a given dose.

Many other projects would benefit from improved detectors, particularly when time resolution is required or, more modestly, for specimen assessment prior to a full experiment. Many of the future applications of x-rays will depend on the ingenuity and level of resources available to detector development teams.

Applications

Of course, the development of new and better sources and detectors as well as x-ray optics, which will be mentioned later, is driven by the applications to which x-rays can be put. Advances in technology not only lead to improved performances for established applications but also allow new uses of x-rays to be envisaged and implemented.

I will consider three areas of study: crystalline materials, non-crystalline materials and lithography.

There can be complete confidence in a secure future for x-rays as probes of the atomic and molecular structures of crystalline materials. The notion prevalent, twenty and more years ago, that the 'phase problem' was a serious barrier has now more or less disappeared. There is no omnicompetent solution to the phase problem but there are sample dependent solutions for most specimens which yield x-ray diffraction data.

X-ray diffraction patterns from crystalline materials with modest molecular weights can be solved by collecting the Bragg intensity data and applying the structural constraint of atomicity. Crystals of molecules with large molecular weights, such as biological macromolecules, can be solved by a variety of techniques such as multiple isomorphous replacement, wavelength variation with anomalous scattering, different crystal forms, molecular replacement and solvation effects.

The current situation is that when good intensity data can be collected in high resolution diffraction patterns from crystals then a rich yield of information is obtainable on atomic and molecular structure.

One can predict with a high degree of confidence that x-ray diffraction will, through the availability of the new brilliant microbeam sources described above, be a powerful tool in the investigation of atomic and molecular structures in low volume specimens such as microcrystals, single magnetic domains, single biological fibres (such as muscle) and thin films and membranes.

The brilliant microbeams will also allow both spatial resolution (particularly in the investigation of polycomposite materials when, once again, biomaterials and biotissues will be fruitful objects of study) and temporal resolution of structures. One can look forward to movies of crossbridge movement during muscle contraction and of other motile systems in living things. Many of these materials are partially crystalline—they are by no means completely disordered. However, full crystallinity would be incompatible with the function of many biological systems and a clear determination of the nature of the deviations from crystallinity and the relationship of these to function are of importance. It is significant that key biological fibres, the diffraction patterns of which seemed intransigent to solution, are now yielding to synchrotron radiation sources.

We can conclude that a safe prediction is that spatially and temporally resolved structure determination of crystalline and partially crystalline materials will be a busy industry, certainly for many decades to come.

The brilliant x-ray sources can also be expected to produce observable diffraction patterns from noncrystalline materials, including single complicated structures such as molecular complexes or even single biological cells of weight 10^{12} daltons. The phase problem can be addressed, as David Sayre has argued[2], by an extension of the general approach described for crystals, particularly those with large repeating units. The advantage for nonperiodic structures is that the continuous Fourier transform can be sampled in the diffraction pattern and this, together with a constraint such as an envelope perimeter, can provide structural information. These experiments at present are being developed using synchrotron radiation with long wavelengths of around 2 nm.

Of course, the analogue means of solving the phase problem is by using a lens and constructing an x-ray microscope. The key elements are a brilliant x-ray source and a focusing optic, such as a Fresnel zone plate, constructed to high fidelity. Techniques for producing such zone plates have been under development in the USA (at IBM and the Center for X-Ray Optics, Berkeley), in Germany (at Göttingen University) and at King's College London. Tim Weiss and I, working with the team from King's College, have reported the imaging of the 65 nm periodicity in collagen[3]. Since we can predict that the advances in nanotechnology will lead to better zone plates there is a strong likelihood that x-ray microscopy will allow imaging of biological cells *in vivo*. Hence, time resolved studies on problems in cell biology which require observations at, say, 10 nm resolution will be possible. One can imagine studies on the life history of the nucleolus or of mitochondria or the Golgi apparatus.

The essence of these possibilities is the existence of a brilliant micrometre cross section x-ray beam. This is already well within prospect by the use of Bragg-Fresnel optics, which are combinations of zone plates and multilayer x-ray mirrors or crystals, coupled with a high-β undulator beam. Tim Weiss and I have shown this in collaboration with Anatoly Snigirev and Christian Riekel at the European Synchrotron Radiation Facility[4], and it is already being applied to spatially resolved studies on mineralising collagen.

The future of x-ray applications is by no means restricted to biology. One example, briefly, which follows from the advances discussed above, is the use of microbeams for microlithography. It may well be that microbeams at synchrotron sources will be used to write the optical elements which are then used to exploit the beams further. Microlithography of three-dimensional objects is also imaginable and, indeed, has been used in the manufacture of micro-machines in recent times[5].

Summary

Whilst it is impossible to predict the specific future for x-rays in basic research, it is possible to be completely confident that, whatever future there is for studies of the fundamental nature of our world, we will continue to see x-rays as playing a central rôle.

References

(1) C Freeman "Economics of Research and Development" *Science, Technology and Society: a Cross-Disciplinary Perspective* (I Spiegel-Rösing & D Price, eds) London: SAGE Publications p 223 (1977)

(2) D Sayre *et al* "On the Possibility of Imaging Microstructures by Soft X-Ray Diffraction Pattern Analysis" *X-Ray Microscopy* (G Schmahl & D Rudolph, eds) Berlin: Springer p 314 (1984)

(3) R E Burge *et al* "X-Ray Microscopy at Suboptical Resolution; Direct Observation of the 65 nm Periodicity in Collagen Fibrils" *J. X-Ray Sci. Technol.* **3** 311 (1992)

(4) A A Snigirev *et al* "A Double Focusing Camera with Micrometre Focal Spot Based on a Circular Bragg-Fresnel Lens" *J. Physique IV* **3**(C8) 443 (1993)

(5) P Bley "The LIGA Process for Fabrication of 3-Dimensional Microscale Structures" *Interdisciplinary Science Reviews* **18** 267 (1993)

Glossary

Cross references to other entries in the glossary are indicated by *italics*.

Ablation. The removal of material from a surface by vaporisation.

Absorbed dose. The energy absorbed by a patient or sample, nowadays measured in *gray*.

Absorption. The transformation of radiation energy into a different form. X-rays are absorbed when they transfer their energy to an electron of an atom or *ion*, removing the electron through the process of *ionisation*.

Absorption coefficient. See *linear absorption coefficient* and *mass absorption coefficient*.

Absorption edge. A discontinuity in the variation of the *linear absorption coefficient* of an element, which arises when an x-ray *photon* has sufficient energy to remove an electron from a particular atomic shell (e.g., *K shell*, *L shell* or *M shell*) and is thus more likely to to be absorbed.

Absorption length. The distance over which the *intensity* of a beam of radiation is reduced by *absorption* to about one third of its initial value.

Accretion. The accumulation of matter by a star through gravitational attraction.

Adiabatic cooling. An adiabatic process takes place without heat exchange with the surroundings. Adiabatic expansion of a gas results in cooling.

Amplified spontaneous emission (ASE). The process by which radiation gains in *intensity* in x-ray lasers. The *spontaneous emission* from *ions* in a *plasma* is amplified by the same type of ion.

Amplitude. The maximum value of a periodically varying quantity.

Ångstrom (Å). A unit of length commonly used to specify x-ray *wavelength*; $1 \text{ Å} = 10^{-10} \text{ m} = 0.1 \text{ nm}$.

Anode. The electrode (conductor) of an x-ray tube held at positive potential with respect to the *cathode*.

Apogee. The furthest point from the Earth in the orbit of a satellite.

Arteriography. The examination of arteries by *radiography* following injection of a contrast medium. The image obtained is called an arteriogram.

Atomic force microscope (AFM). An instrument in which the electrons in a fine metal probe are repelled by those of the atoms on the surface of a sample. As the probe is scanned across the surface its height is adjusted to keep the force on it constant. A sensing mechanism records the up and down movements of the probe, giving an image of the surface at atomic resolution.

Atomic number. A number defining the order of the elements in the periodic table; equal to the number of protons in the nucleus.

Atomic weight. The mass of an atom measured in atomic mass units (amu); 12 amu is equal to the atomic weight of ^{12}C, the most common form of carbon.

Avogadro's number. The number of molecules in one *mole* of any substance; $N_A=6.023\times10^{23}$ mol^{-1}.

Beamline. A collection of optical components designed to transmit x-rays from the source (usually a *storage ring*) to an experiment.

Binding energy. The energy required to free an electron from a *bound state*.

Blackbody. A body which totally absorbs incident *electromagnetic radiation*. The radiation emitted from an ideal blackbody has an energy distribution which depends only on its temperature.

Black hole. An astronomical object with a gravitational field so high that neither particles nor *photons* can escape.

Bound-bound transition. A transition between two *bound states* of an atom or an *ion*. An electron in an *excited state* will relax to a state of lower energy through emission of *electromagnetic radiation* of a discrete energy.

Bound state. A state of a system, such as an electron in an atom or an *ion*, which has a discrete energy. See also *ground state* and *excited state*.

Bremsstrahlung. *Electromagnetic radiation* emitted when a charged particle (usually an electron) is decelerated as the result of a close approach to the nucleus of an atom or *ion*. The continuous *spectrum* of an x-ray tube is produced by bremsstrahlung. See also *free-free transition*.

Brewster angle. The angle of incidence at which *electromagnetic radiation* reflected from a surface is plane polarised (see *polarisation*). The tangent of the Brewster angle is equal to the *refractive index* of the surface material.

Brightness. The radiation *flux* per unit *solid angle* of emission.

Brilliance. The radiation *flux* per unit area per unit *solid angle* of emission.

Cathode. The electrode (conductor) of an x-ray tube which emits a stream of electrons.

Cathode rays. The stream of electrons emitted from a *cathode*.

Cation. A positively charged *ion*.

Chemical bond. The electronic force holding atoms together in molecules.

Chirped pulse amplification. A technique to produce ultrashort laser pulses of exceedingly high peak powers (terawatts) by amplifying and then compressing low *intensity* pulses.

Code. A computer program used to simulate the behaviour of a *plasma*.

Coherence. Two waves with constant *phase* relationship are said to be coherent.

Collimator. A device for obtaining a parallel beam of radiation.

Collisional radiative equilibrium. A description of the state of a *plasma* with density between those for which *local thermodynamic equilibrium* and *coronal equilibrium* apply.

Corona. The outer part of the Sun's atmosphere, a tenuous *plasma* with a temperature of about one million degrees *kelvin* (10^6 K).

Coronal equilibrium. A description of the state of a *plasma* with properties similar to those of the solar *corona*.

Covalent bond. A *chemical bond* in which atoms are held together by interaction of their nuclei with electron pairs.

Covalent radius. Half the separation distance between the nuclei of two atoms held together by a *covalent bond*.

Critical angle. The *grazing angle* below which a material can have a high reflectivity for x-rays. Critical angles are typically a few degrees.

Critical density. The number of electrons per cubic metre in a *plasma* above which *electromagnetic radiation* of a certain *frequency* cannot propagate. The critical density is proportional to the square of the frequency.

Crystal. A solid which has its atoms or molecules arranged in a systematic geometrical pattern.

Curie (Ci). A unit of radioactive decay; 1 Ci$=3.7\times10^{10}$ decays per second.

Densitometry. The measurement of the optical density of exposed film.

Diffraction. The spreading of *electromagnetic radiation* beyond the geometrical shadow of an obstruction. Diffraction becomes noticeable when the size of the obstruction is comparable to the *wavelength* of the radiation.

Diffraction grating. In its simplest form, a series of parallel bars and spaces of equal width which cause *electromagnetic radiation* to be deviated by *diffraction* through angles depending on the *wavelength*. Hence diffraction gratings may be used to measure wavelength.

Diffraction limited beam. A beam of radiation whose diameter is limited only by the effects of *diffraction*.

Diffraction pattern. A pattern formed by equal *intensity* contours as a result of the *diffraction* of *electromagnetic radiation* by an object.

Divergence. The increase in cross sectional area of a beam of radiation or charged particles.

DNA. Deoxyribose nucleic acid, the genetic material of organisms.

Doppler effect. The apparent change in *frequency* (or *wavelength*) of radiation from a moving source. The magnitude of the change is the Doppler shift.

Dose. A quantity of radiation. See *absorbed dose*.

Electric field. The region surrounding an electric charge within which it may exert a force on other electric charges.

Electromagnetic radiation. Waves, caused by the acceleration of charged particles, consisting of *electric fields* and *magnetic fields* vibrating transverse to the direction of motion and at right angles to one another.

Electronvolt (eV). The energy acquired by an electron when it passes through an electrical potential difference of one volt; 1 eV$=1.602\times10^{-19}$ J.

Emittance. The product of the beam size and angular *divergence* of *synchrotron radiation*.

Energy level. An allowed value of energy which, for example, an electron can have when it is in a *bound state* of an atom, an *ion* or a molecule.

Epitaxial growth. A method of growing successive thin layers of crystalline material with the *crystal* orientation maintained from layer to layer.

Equation of state. An equation relating the physical characteristics of a system.

Excited state. The state of a system, such as an atom or an *ion*, with an energy higher than its *ground state*.

Femto (f). The prefix for 10^{-15} of a base unit; e.g., $1 \text{ fs} = 10^{-15}$ s.

Fluorescent radiation. The *electromagnetic radiation* emitted by certain substances when struck by radiation of higher energy.

Fluoroscopy. The examination of an object by viewing its shadow on a screen which emits *fluorescent radiation* when excited by x-rays (a fluoroscope).

Flux. The rate of flow of radiation energy across a unit area perpendicular to the direction of propagation.

Focus. A point to which rays converge after passing through an optical system.

Fourier transform. A mathematical operation which determines the *harmonic* components of a complex waveform.

Fraunhofer diffraction. The *diffraction* produced when an object is illuminated by a parallel beam of radiation. The *diffraction pattern* is observed at a large distance from the diffracting object.

Free-bound transition. A transition in which a free electron is captured into a *bound state* of an atom or an *ion*, resulting in the emission of *electromagnetic radiation* (recombination radiation) over a wide range of *wavelengths*.

Free electron laser (FEL). A laser utilising a beam of electrons spiralling around *magnetic field* lines, for example in an *undulator* of a *storage ring*. The laser *wavelength* can be tuned by changing the electron beam energy or the magnetic field.

Free-free transition. A transition between two free electron states resulting in the emission (*bremsstrahlung*) or *absorption* (*inverse bremsstrahlung*) of *electromagnetic radiation*.

Fresnel diffraction. The *diffraction* produced when an object is illuminated by an expanding beam of radiation from a point source.

Frequency. The number of times that a regularly recurring phenomenon occurs in one second. An x-ray of *wavelength* 1 nm corresponds to a frequency of about 3×10^{18} Hz, where 1 Hz (hertz) is one cycle per second.

Fundamental. The *harmonic* of lowest *frequency* in a waveform.

Fusion. The formation of elements from others of lower *atomic number*; in fusion reactors helium is formed from isotopes of hydrogen, releasing energy.

Gamma rays or γ-rays. The *electromagnetic radiation*, of shorter *wavelength* than x-rays, produced by transitions between the *energy levels* of excited atomic nuclei.

Geiger counter. An instrument that detects the passage of ionising radiation through a gas filled tube in which a strong *electric field* is maintained by a wire at high potential.

Giga (G). The prefix for a billion of a base unit; e.g., $1 \text{ GeV} = 10^6 \text{ eV}$ (*electronvolts*).

Gray (GY). The unit of *absorbed dose*; $1 \text{ GY} = 1 \text{ J kg}^{-1}$.

Grazing angle. The incidence angle a beam of radiation makes with the surface of a reflector.

Ground state. The state of a system, e.g., an atom or an *ion*, at its lowest energy.

Hæmoglobin. The constituent of blood which carries oxygen.

Harmonic. A component of a waveform with a *frequency* which is an exact multiple of the *fundamental.*

Holography. A technique for registering the *phase* as well as the *amplitude* of the radiation emanating from an illuminated object by recording the *interference* pattern (hologram) of this radiation with a reference beam. This allows the production of three-dimensional images.

Hydrogen bond. A strong *chemical bond* found in liquids, such as water, containing hydrogen.

Image contrast. A measure of the difference between the brightest and darkest parts of an image.

Infrared radiation. *Electromagnetic radiation* just beyond the limit of red visible light. The *wavelengths* of infrared radiation (about 1 µm to 1 mm) are shorter than those of radiowaves and longer than those of light waves.

Insertion device. See *undulator* and *wiggler.*

Intensity. The rate of energy transfer through a unit area perpendicular to the propagation direction of a beam of radiation.

Interference. An effect which occurs when two or more waves combine to give a resulting wave whose *amplitude* depends on the *frequencies*, relative *phases* and amplitudes of the interfering waves.

Interferometer. An instrument that uses the *interference* of radiation for precise measurement of *wavelength*, small distances and optical phenomena.

Inverse bremsstrahlung. A process in which a *photon* and an electron collide simultaneously with an *ion*, the photon is absorbed and its energy is transferred to the electron.

Ion. An atom which has had some (or all) of its electrons removed, and thus has an overall positive charge.

Ionisation. The process by which electrons are removed from atoms to form *ions*. The energy required for ionisation—the ionisation potential—can be provided by collisions or by radiation.

Ionosphere. The part of the Earth's upper atmosphere which consists of ionised material, i.e., of electrons and *ions.*

Irradiance. The *flux* of *electromagnetic radiation* incident per unit area on a surface.

Joule (J). A unit of energy equivalent to the work done when a force of $1 \, kg \, m^{-1} s^{-2}$ moves its point of application through 1 m.

Kelvin (K). A temperature scale in which absolute zero, 0 K, is at -273.15° C. The freezing and boiling points of water are at 273.15 K and 373.15 K, respectively, and the degree intervals are identical to those of the Celsius scale.

Kilo (k). The prefix for a thousand of a base unit; e.g., 1 kpc=10^3 pc (*parsecs*).

Kinetic energy. The energy associated with motion.

KrF laser. A type of laser in which a mixture of krypton and fluorine gases known as an excimer is used as the lasing medium. The *wavelength* of the laser radiation is 249 nm. Similar lasers can be made from combinations of other rare gases (e.g., argon, xenon) with halogens (e.g., chlorine).

K shell. The innermost *energy level* of an atom or *ion*, corresponding to a *principal quantum number* of $n=1$, and containing up to two electrons. A *bound-bound transition* to the K shell results in characteristic K radiation.

Least squares procedure. A method of fitting a curve to a trend in data.

LIGA. A process for making submillimetre sized machines or components by x-ray *lithography*; derived from a German acronym Lithographie, Galvanoformung (electroplating) and Abformung (moulding).

Light curve. The change in *brightness* with time of an astronomical object.

Linear absorption coefficient. When a beam of radiation passes through a thickness t of a material the incident *intensity* is reduced by a factor $\exp(-\alpha t)$ where α is the linear absorption coefficient.

Lithography. A method of copying a master pattern. Advanced lithographic techniques, e.g., optical, x-ray and electron beam lithographies, are used in the microfabrication of integrated circuits and other semiconductor devices.

Local thermodynamic equilibrium (LTE). A description of the state of a *plasma* in which the *excited state* populations are determined solely by collisions.

L shell. The *energy level* of an atom or *ion* with a *principal quantum number* of $n=2$, containing up to eight electrons. A *bound-bound transition* to the L shell results in characteristic L radiation.

Mach-Zehnder interferometer. A type of *interferometer* in which the *interference* fringes can be localised in a chosen plane.

Macromolecule. A very large molecule, containing thousands of atoms, or a high molecular mass *polymer*.

Magnetic field. The area of force surrounding a magnetic pole or a conductor in which an electric current is flowing.

Mass absorption coefficient. The ratio of the *linear absorption coefficient* and the density of a material, commonly used to describe the x-ray *absorption* properties of elements and compounds.

Matrix composition. The composition of the structure containing an element or compound which is to be analysed.

Mega (M). The prefix for a million of a base unit; e.g., $1\,Mpc = 10^6\,pc$ (*parsec*).

Metrology. The science of measurement. More specifically, the measurement of a component to assess its actual performance compared to that required.

Micro (μ). The prefix for one millionth of a base unit; e.g., $1\,\mu m = 10^{-6}\,m$.

Milli (m). The prefix for one thousandth of a base unit; e.g., $1\,ms = 10^{-3}\,s$.

Millibar (mbar). A unit of pressure; 1000 mbar, or 1 bar, is approximately equal to standard atmospheric pressure.

Mole. The amount of a substance that contains as many entities (e.g., atoms, molecules) as there are in 12 g of ^{12}C.

Molecular weight. The sum of the *atomic weights* of all the atoms contained in a molecule.

Monochromator. A device used to select a small *wavelength* range from *electromagnetic radiation* with a continuous *spectrum*.

Monomer. A unit from which a *polymer* is built.

M shell. The *energy level* of an atom or *ion* with a *principal quantum number* of $n=3$, containing up to eighteen electrons. A *bound-bound transition* to the M shell results in characteristic M radiation.

Multilayer mirror. A mirror made of successive layers of material arranged so that *reflections* from the interfaces between each layer add in *phase* to enhance the overall reflectivity.

Mylar. The tradename for a type of plastic film used, e.g., as the base for audiotape and videotape.

Nano (n). The prefix for one billionth of a base unit; e.g., $1 \text{ nm} = 10^{-9}$ m.

Neutron star. A star which, after gravitational collapse as the result of a *supernova* explosion, has a high enough density to make it energetically favourable for all its protons to convert into neutrons by capturing electrons.

Nova. An evolved star which suddenly brightens by a factor of 10 000 or more.

Opacity. The ability of a material to absorb *electromagnetic radiation.*

Parsec (pc). The distance to a star that shows an angular shift of two seconds of arc when viewed from opposite sides of the Earth's orbit; $1 \text{ pc} = 30.86 \times 10^{12}$ km.

Peta (P). The prefix for a million billion of a base unit; e.g., $1 \text{ PW} = 10^{15}$ W.

Phase. The elapsed fraction of a cycle of a periodic waveform. One cycle corresponds to movement through an angle of 2π *radians*, i.e., 360°.

Phase velocity. The velocity of the nodes of a travelling wave.

Photon. A *quantum* of *electromagnetic radiation.* Radiation sometimes behaves as waves and sometimes as a stream of small quantities, or quanta, of energy. The energy, E, of a photon is given by $E=h\nu$, where ν is the *frequency* of the radiation and h is Planck's constant ($h=6.626 \times 10^{-34}$ J s $=4.136 \times 10^{-15}$ eV s).

Photoresist. A *polymer* which, when exposed to radiation, becomes either easier (positive photoresist) or harder (negative photoresist) to dissolve.

Photosphere. The layer of hot gases nearest to the Sun's surface.

Pico (p). The prefix for 10^{-12} of a base unit; e.g., $1 \text{ ps} = 10^{-12}$ s.

Planck's constant. See *photon.*

Plasma. A state of matter in which some or all of the atoms or molecules are ionised to give a collection of *ions* and electrons.

Polarisation. Polarised radiation consists of *photons* whose *electric fields* are all aligned in the same direction.

Polarising angle. See *Brewster angle.*

Polymer. A substance built up from many identical smaller units (*monomers*).

Population inversion. The required environment for laser action, in which an *energy level* of a system is made to contain relatively more particles than a level of lower energy.

Principal quantum number. An integer which defines the *energy level* of an electron bound in an atom or *ion.* For hydrogen, and hydrogen like ions, the energy of the level with principal quantum number n is proportional to $-1/n^2$, where the negative sign indicates a *bound state.*

Proportional counter. A detector, similar to a *Geiger counter*, which gives an output pulse proportional to the *ionisation* caused by the passage of radiation.

Q-switch. A device which, in one state, allows the *population inversion* of a laser to build up and, in the other state, allows rapid depopulation, resulting in pulsed laser output of very high power.

Quantum. An indivisible unit of energy, such as the *photon* for *electromagnetic radiation*.

Quantum mechanical tunnelling. The phenomenon, forbidden in classical physics, in which a particle passes through an energy barrier larger than the energy of the particle.

Quasar. A distant, compact and very bright astronomical object, whose radiation output indicates that it is probably powered by a *black hole*.

Radian (rad). A measure of angle; 2π rad$=360°$, i.e., 1 rad$\approx57°$.

Radiography. The recording of shadow images (radiographs) using x-rays.

Radiology. The branch of medicine concerning the use of x-rays and other forms of radiation in the diagnosis and treatment of disease.

Raman shift. The change in *frequency* of *electromagnetic radiation* when *scattered* by molecules which gain or lose energy via transitions between *energy levels* associated with the vibration and rotation of the molecules.

Reciprocal space. A hypothetical space in which the reciprocal lattice, a mathematical aid for describing *diffraction patterns*, can be considered to exist.

Recombination radiation. See *free-bound transition*.

Redshift. The displacement of spectral features to longer *wavelengths* caused, e.g., by the *Doppler effect*.

Reflection. The deflection of incident radiation by the surface of a material. The reflected *intensity* depends on the *refractive index* of the material.

Refraction. The change in direction of a beam of radiation as it crosses the boundary between two media with different *refractive indices*.

Refractive index. A property of a material equal to the ratio the speed of light in a vacuum to the *phase velocity* of *electromagnetic radiation* in the material.

Relativistic effects. The phenomena, such as increase in mass, contraction of length and time dilation, noticeable at speeds close to that of light (3×10^8 m s^{-1}).

Resist. See *photoresist*.

Resolving power. The ability of an imaging system to distinguish closely spaced objects or, for a spectroscopic system, closely spaced *wavelengths*.

Resonance transition. A transition between two states of a *plasma* caused by incident radiation and resulting in the emission of resonance radiation.

Roentgen. A unit of x-ray *dose* equivalent to the liberation, through *ionisation*, of 1.61×10^{15} electron charges per kilogram of air.

Scattering. The deflection of radiation from the main direction of a beam.

Scintillator. A material which emits flashes of light when struck by ionising radiation such as x-rays.

Secondary electrons. The electrons emitted from a surface when it is bombarded by other electrons.

Seyfert galaxy. A galaxy, resembling a *quasar*, with a brilliant nucleus and faint spiral arms.

Signal to noise ratio. The ratio of the signal in, e.g., an image to the level of background noise. If this is too low *image contrast* is adversely affected.

Skiagraphy. An archaic term for *radiography*.

Solid angle. The ratio of the observed area of an object to the square of the distance from the point of observation. See *steradian*.

Spectrum. The distribution of *electromagnetic radiation*, as a function of its *frequency* or *wavelength*, which includes (in order of increasing wavelength) γ-rays, x-rays, *ultraviolet radiation*, visible light, *infrared radiation*, microwaves and radiowaves.

Specular reflection. *Reflection* in which the angle of the reflected radiation, with respect to the reflecting surface, is equal to that of the incident beam.

Spontaneous emission. The emission of *bound-bound* radiation from an atom or *ion* without external stimulation.

Standard. A reference component, measurement or instrument which is used in comparing results from measurements on a range of similar systems.

Steradian (sr). The unit of *solid angle*. A sphere subtends a solid angle of 4π sr at its centre.

Stereochemistry. The arrangement of atoms within a molecule.

Stimulated emission. A process in which an incoming *photon* of a particular *frequency* induces an atom or *ion* to emit another photon of the same frequency and *phase*.

Storage ring. A roughly circular vacuum pipe in which beams of charged particles are kept orbiting by *magnetic fields*.

Supernova. A star undergoing gravitational collapse, causing a great increase in *brightness* and the formation of a *neutron star* or, possibly, a *black hole*.

Synchrotron radiation. The radiation emitted by high energy charged particles, particularly electrons, when they are made to move in curved paths by *magnetic fields* in, e.g., *storage rings*, synchrotrons and some astronomical objects.

Tera (T). The prefix for a thousand billion of a base unit; e.g., $1\,TW=10^{12}\,W$.

Tokamak. A toroidal apparatus used in *fusion* research for confining a *plasma* by two *magnetic fields*.

Total external reflection. The *reflection* of x-rays incident on a surface at a *grazing angle* below the *critical angle* when, for a hypothetical nonabsorbing material, all the x-rays would be reflected.

Ultraviolet radiation. *Electromagnetic radiation* between the visible and x-ray parts of the *spectrum*, with *wavelengths* from about 5 nm to about 400 nm. Radiation with wavelength below about 200 nm, which is absorbed by air, is referred to as the vacuum ultraviolet.

Undulator. An array of magnets, placed in a straight section of a *storage ring* which causes the electron beam to take an oscillatory path. X-rays emitted from points on the path with the same *phase* add coherently to enhance the *brilliance* of the radiation.

Valence electrons. Electrons, in the outer shells of atoms, which are involved in chemical reactions.

Vector. A quantity, such as velocity, which has both magnitude and direction.

Wavelength. The distance between two points of a wave which have the same *phase*. X-rays have wavelengths in the range (roughly) of 0.1 to 10 nm.

Wiggler. An array of magnets, placed in a straight section of a *storage ring* which causes the electron beam to travel in a sharper curve and enhances the x-ray *flux*.

X-radiation. The same as x-rays.

XUV radiation. A term used to describe the region of the *spectrum* encompassing x-rays and vacuum *ultraviolet radiation*.

YAG laser. A type of laser in which the lasing medium is glass containing neodymium in yttrium aluminium garnet. The *wavelength* of the laser radiation is 1.064 µm.

Zone plate. A type of *diffraction grating* in which the period decreases in a defined way outwards from the centre. Zone plates are usually circular and are used to *focus* x-rays using *diffraction*.

Index

Page numbers in *italics* refer to glossary entries.

Ablation, 194, 206, *249*
Absorption, 6–7, 71, 72, 162–3, 200–1, 203, 239, 245, *249*
  coefficient, 61–2, *254*
  edge, 49, 56, 90, 148, 151, 245, *249*
  length, 49, *249*
  spectroscopy, 9, 199, 202–3
Accretion, 187, 188–9, *249*
Advanced Light Source (Berkeley), 55
American Roentgen Ray Society, 34
Amplified spontaneous emission, 210, 227, 235, *249*
Arteriography, 33, *249*
Astigmatism, 50
Astronomy, 12, 175–90
  rocket experiments, 12, 176, 178–9
  satellite experiments, 176, 177, 179–86, 190
Atomic energy levels, 5, 64, 197, 207, 209, *251*
Avogadro's number, 11, *250*

Balmer series, 195, 209, 235
Blackbody emission, 194, 206, 208, *250*
Bragg's law, 7, 8, 63
Bremsstrahlung, 4–5, 12, 65, 194, 225, *250*
  inverse, 194, *253*
Brewster angle, 214, *250*

Cathode rays, 2, *250*
Cavendish Laboratory (Cambridge), 14, 43, 44, 47, 48, 80
Center for X-Ray Optics (Berkeley), 247

Central Laser Facility (Rutherford Appleton Laboratory), 13, 204, 212, 222, 228, 229, 240
Chirped pulse amplification, 219–220, 229, 235–6, *250*
Collimation, 34, 139, 150, 178, *250*
Critical angle, 74, *251*
Crystal, 3, 7, 116, 123, 246, *251*
  lattice, 7, 11, 106, 107–8, 199
  spectroscopy, 4, 63
  structure, 7, 105–6, 109, 121
Crystallography, 4, 8, 110, 147, 150
  protein, 116, 128, 243

Detectors
  charge coupled device, 48, 102, 186, 189, 239
  Geiger counter, 67, 176, *252*
  photographic film, 7, 33, 39, 176, 228
  photographic plate, 1, 3, 12, 43, 63
  photoresist, 43, 57, 204, *see also* Photoresist
  proportional counter, 67, 68, 84–5, 176, 179, 182, 184, *255*
  scintillation counter, 67, 68
  solid-state, 85, 86–7, 95
  streak camera, 199
Diffraction, 3, 51, 63, 101–28, 148, 246–7, *251*
  crystal, 7–8
  Fraunhofer, 102–4, *252*
  function determination by, 113, 118, 125–6, 147–8
  grating, 8, 51, 215, *251*
  Laue, 127
  pattern, 7, 102–8, 110–11, 113, 123, 247, *251*

phase problem, 148, 246, 247
  direct methods, 109–10
  heavy atom method, 108–9
  isomorphous replacement, 109
  Patterson synthesis, 104, 108
  structure determination by, 8, 104,
    108–9, 111, 113, 115–17, 121–3,
    147, 246
DNA, 56, 106, 110–11, *251*
  double helix, 8, 106, 108, 111

Electronic area detector, 48, 102, 186,
  189, 239
Elemental mapping, 48, 55–6, 76,
  80–2
Electron probe microanalysis
  (EPMA), 9, 61–96
Equilibrium
  collisional radiative, 207, *250*
  coronal, 207, *250*
  local thermodynamic, 202, 206, *254*
European Synchrotron Radiation
  Facility (Grenoble), 55, 123,
  152–3
EXAFS, 90, 148, 150, 151, 199

Fluorescence, 9, 66–75, 89, 90, 178,
  189, *252*
Fluoroscopy, 26, 28, 31, 33, *252*
Four wave mixing, 220
Fourier synthesis, 104, 108–9,
  116–17
Fourier transform, 104, 105, 106, 247,
  252
Free electron laser, 144–5, 154, *252*
Free-bound transition, 194, 195, *252*
Free-free transition, 194, *252*, *see also*
  Bremsstrahlung
Fusion, *252*
  inertial confinement (laser induced),
    12–13, 199, 205–6, 226–7, 239
  magnetic confinement (Tokamak),
    198, *257*

Harmonic generation, 220, 231–3
Helios, 164–70
Hohlraum, 205–6
Holography, *see* X-ray holography

Interference, 143, 215, *253*
Interferometry, 11, 223, *253*, *254*
Ionisation, 2, 5, 64, 175–6, 193,
  194–5, 232, *253*
Irradiance, 194, 198, 212, *253*

King's College London, 8, 16, 21, 30,
  54, 55, 247

Laser, *see* X-ray laser
Lawrence Livermore National
  Laboratory, 13, 211, 219, 222,
  223
Lithography, *254*
  optical, 157–61
  x-ray, 9–10, 14, 155–72, 204–5, 248
Lyman series, 195

Microanalysis, 9, 61–96
Microscopy
  electron, 48, 49, 79–80, 84
  x-ray, 43–58, 247
    contact, 7, 43–4, 76, 203–4, 221
    image contrast in, 49, 203, *253*
    magnification in, 43, 46
    projection, 14, 44–8
    reflection, 14, 44
    scanning, 54–6
    transmission, 14, 53–4
Monochromator, 138, 139, *254*
Multilayer mirror, 18, 50, 139, 210,
  223, *255*

National Synchrotron Light Source
  (Brookhaven), 55
Naval Research Laboratory
  (Washington), 12, 176, 212
Nobel Prize, 1, 24

Optics, 14–18, 51
  Bragg-Fresnel, 248
  Kirkpatrick-Baez, 14
  multilayer mirror, 18, 50, 139, 210,
    223, *255*
  Wolter, 16
  zone plate, 15, 51–3, 247, *258*

Photoresist, 9, 158, 162, *255*, *see also* Detectors, photoresist
Plasma, 175–6, 193–223, *255*, *see also* Sources, plasma
  critical density, 236, *251*
  density, 194, 202, 212, 217
  temperature, 176, 194, 203, 208, 212
Population inversion, 207, 209, 227, *255*
Proton excited x-ray emission, 90–4

Radiation
  characteristic, 5–6, 61–4, 195–8, 225–6, *254*, *255*
  continuous, 4–5, 10, 65, 225, *see also* Bremsstrahlung
  dose, 38–9, 245, *249*
  electromagnetic, 3–4, 64, 193, 225, *251*
  protection, 34
  recombination, 194, 195, *256*
  synchrotron, *see* Synchrotron radiation
  units, 38–9
Radiography, 6–7, 23, 26–9, 33, 199, 238–9, *256*
Reciprocal lattice, 106, 107–8
Reciprocal space, 105, *256*
Reflection, 17–18, 44, 50, 139, *256*
  grazing incidence, 8, 14, 50, *252*
  total external, 8, 14, 18, 74, *257*
Refraction, 49–50, 212–13, 228–9, *256*
Refractive index, 8, 14, 49–50, 233, *256*
Resolution (resolving power), *256*
  angular, 16, 183
  spatial, 16, 43, 52, 53, 103, 228
  spectral, 138, 186, 189, 195
Resonance transition, 195, 197, 201, *256*
Röntgen rays, 2, 24, 25, 190
Röntgen Society of London, 21, 30, 34

Scattering, 3, 108–10, 148, 178, *256*
  anomalous, 113, 117, 246
Solar corona, 175–6, 199, *250*

Solar flare, 176, 199
Solar photosphere, 175, 176, *255*
Sources
  extraterrestrial, 12, 175–90
  plasma, 56, 193, 198–200, 204, *see also* Plasma
  storage ring, 134, 135–6, 138, 143, 152, 164–6, *257*, *see also* Synchrotron
  x-ray tubes, 1–2, 21, 26, 29, 30, 37, 67–8, 225–6
Spectrometer, 62–3, 67–72, 84–7, 95–6
Spectroscopy, 65, 134, 198–9
  absorption, 9, 148, 199, 202–3
  Bragg, 67–70, 72
  energy dispersive, 72–3, 84–7
Spontaneous emission, 225, 227, *257*
  amplified, 210, 227, 235, *249*
Stimulated emission, 225, 227, 235, *257*
Structure factor, 106, 108, 109–10
Synchrotron, 13, 132, 133, 152–4, 166, 226, 244, 245, *see also* Sources, storage ring
  beam lifetime, 168
  beamline, 137–8, 139, 169
  dipole magnet, 141, 142
  electron beam, 136, 139, 226
  insertion device, 141, 143
  undulator, 141–2, 143–4, 238, *257*
  wiggler, 141, 142, *258*
Synchrotron radiation, 126–7, 131–54, 161, 166, *257*
  angular divergence, 136, 168–9, *251*
  brightness, 142, *250*
  brilliance, 135, 141, 143, 145, *250*
  coherence, 143–5, *250*
  critical energy/wavelength, 137, 161–2, 165
  emittance, 135, 143, 244, *251*
  flux, 135, 140, 141, *252*
  spectrum, 13, 132, 136–7, 142, 161, *257*
Synchrotron Radiation Source (Daresbury), 13, 119, 123, 126, 134, 136, 137, 141, 146, 164

Tantalus (Wisconsin), 134

Telescope, 16–17, 183, 185–6
Therapy, 36–7

University of Göttingen, 54, 247
University of Leicester , 12, 176, 177,
    179, 182, 184

Water window, 49, 203, 219
Würzburg University, Physical
    Institute, 1, 21, 22

X-ray absorption edge fine structure
    (XAFS), 90, 148, 150, 151, 199
X-ray holography, 57–8, 221, *253*
X-ray laser, 13, 207–23, 225, 227,
    231–6, 237–9
  amplification, 207, 210, 213–14,
    227, 234–5
  coherence, 210, 215–16, *250*
  collision pumped, 209, 211–19
  gain, 207, 210, 211, 212, 219, 236
  photopumped, 208
  pumping, 207, 208, 210, 217, 219
  recombination, 201, 209, 210, 219
  saturated output, 213–14, 215, 216,
    221, 227, 235
X-ray quality, 40, 61

Zone plate, 15, 51–3, 247, *258*